ILULISSAT ICEFJORD

Editors:
Ole Bennike, Naja Mikkelsen, Henrik Klinge Pedersen and Anker Weidick
Photography:
Jakob Lautrup

GEUS

Editors: Ole Bennike, Naja Mikkelsen, Henrik Klinge Pedersen and Anker Weidick
Journalistic editing: Rolf Haugaard Nielsen
Translation: Michael de Laine and A. K. Higgins
Photography: Jakob Lautrup (unless otherwise indicated)
Cover and lay-out: Henrik Klinge Pedersen
Reprographic work: Benny Schark and Jakob Lautrup
Illustrations: Henrik Klinge Pedersen and others. Original sources are indicated.
Maps: Margareta Christoffersen and Willy L. Weng
3D interpretation: Carsten Thuesen
Satellite interpretation: Tapani Tukiainen

Printing: Schultz
Published: 2004 (2nd reprint, 2006)

ISBN 13: 978 87-7871-136-6

The book can be purchased in **bookshops**, at **GEUS**. Phone +45 38 14 20 00/*geus@geus.dk*,
at **Geografforlaget.** Phone +45 63 44 16 83/*go@geografforlaget.dk* or
at **Atuagkat**, Box 1009, 3900 Nuuk, Phone + 299 32 17 37/*mail@atuagkat.gl*
The book has been published with support from

DANCEA
Danish Cooperation for Environment in the Arctic
Ministry of The Environment

Ilulissat municipality

The book has been published on the occasion of the inscription of the icefjord on UNESCO's World Heritage List

SUNSET OVER THE ICEFJORD

THE SPIRIT IN THE WATER

Contents

ISFJELDSBANKEN ('THE ICEBERG BANK') AT THE MOUTH OF THE ICEFJORD

Preface

LAYER ON LAYER OF ICE

It is a great pleasure to present this work on the Ilulissat Icefjord World Heritage Site. This is the first site in Greenland, and one of the first in the Arctic, to be included on UNESCO's World Heritage List amongst numerous other world famous localities.

The Ilulissat Icefjord site is noted for its unique glaciological characteristics and its scenic beauty. The spectacular icebergs that fill the fjord derive from the glacier at the head of the fjord, Sermeq Kujalleq, one of the fastest moving and most productive glaciers in Greenland. In addition to its direct beauty, the area also contains several of Greenland's most important cultural sites. At the mouth of the icefjord lies the archaeological site Sermermiut, where the earliest settlement was established 4400 years ago. It is the rich marine life of the icefjord that has supported the existence of a succession of cultures for thousands of years, and fishing is still today the most important commercial activity in Ilulissat. In recent years, due to the unique scenic attraction of the area, tourism has grown to become the second most important business activity.

It has been a long and exciting process to obtain recognition of the Ilulissat Icefjord region as a World Heritage List site. This process was the result of exemplary co-operation between representatives from Danish authorities, the Greenland Home Rule authority and Ilulissat municipality. The nomination documentation was completed and submitted in early 2003, and the formal approval of the Ilulissat Isfjord area for inclusion on the prestigious list took place at the 28th meeting of the World Heritage Committee in the summer of 2004.

There is certainly good reason to be proud of the wide recognition that the Ilulissat Icefjord region will receive following its inclusion on the World Heritage List. The Greenland government ('Landsstyret') and Ilulissat municipality will continue with initiatives towards protection and management of this World Heritage site. Because of the increased number of tourists to be expected, the area will be closely monitored in the future, so that coming generations can continue to enjoy the cultural heritage and unspoilt nature of the region.

This book is a popular version of the nomination volume. The parties involved in

Anthon Frederiksen
Mayor of Ilulissat

Jens Napãtôk
'Landsstyre' member

the preparation of the original nomination report agreed to publication in book-form, in Greenlandic, Danish and English versions, of a volume that a wider range of readers can enjoy. This book was prepared by the Geological Survey of Denmark and Greenland (GEUS), and includes many contributions from relevant specialists. It is published with the financial support of the Danish Ministry of the Environment, as part of the environmental support programme Dancea - Danish Co-operation for Environment in the Arctic and Ilulissat municipality.

We hope that those who enjoy reading this book, and admire the varied and fascinating illustrations, will feel tempted to spend a few summer days at Sermermiut, from where the beauty of the icefjord can be properly appreciated.

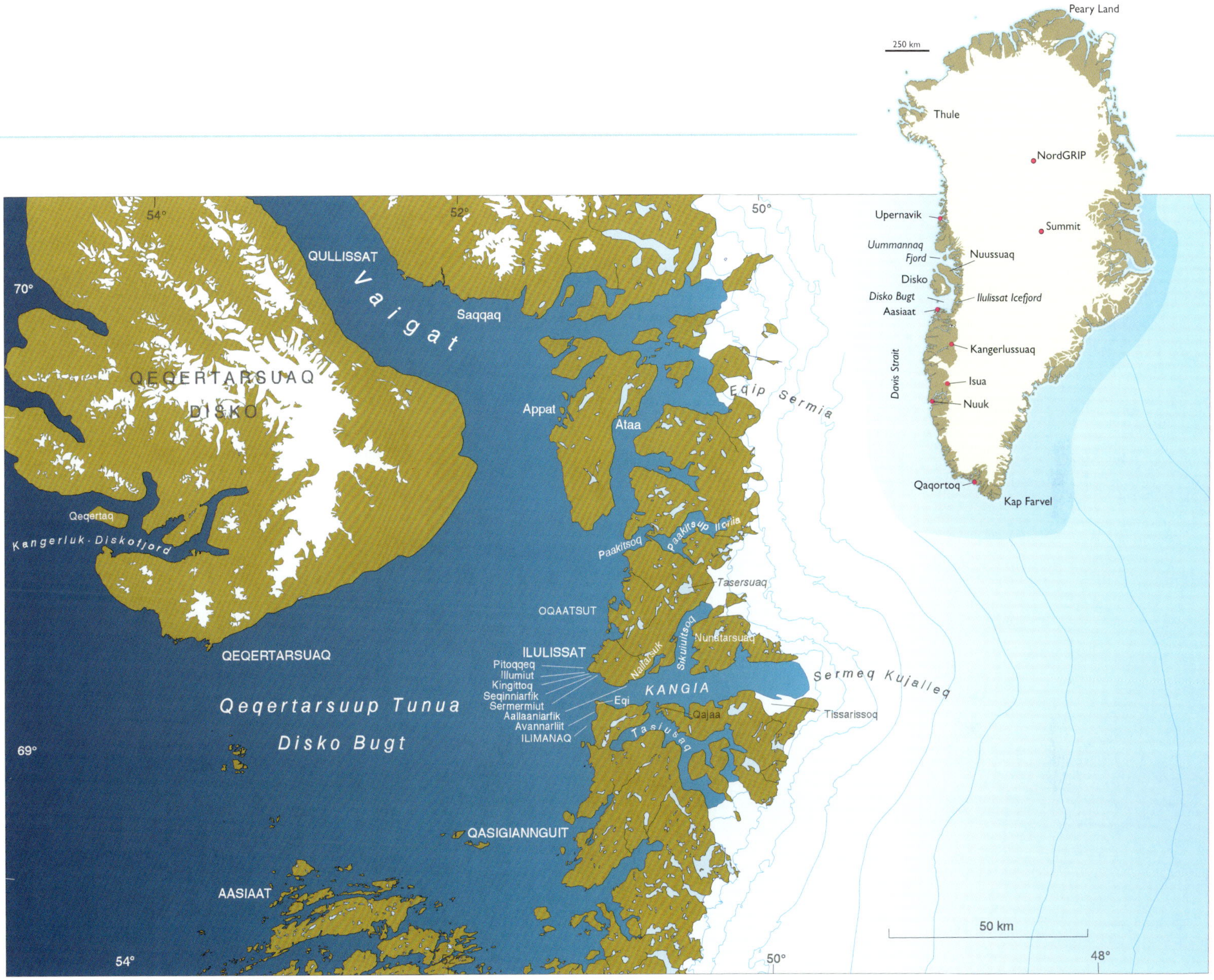

Map showing the Disko Bugt region with place names used. The inset shows a map of Greenland with selected place names.

'HALF A MILE NORTH OF SANT-BAY THERE IS A FJORD, WHICH IS ALWAYS FILLED WITH ICE WITH FEARFULLY HIGH ICEBERGS, BUT WHERE THEY COME FROM SO FAR REMAINS UNKNOWN; THIS FJORD IS CALLED YS-FIOERT'.

FROM THE EARLIEST DESCRIPTION OF KANGIA, PUBLISHED IN 1719 BY THE DUTCH WHALER LOURENS FEYKES HAAN

Ilulissat Icefjord - a unique natural phenomenon

The beautiful and majestic icebergs, that fill the fjord and strand on the bank south-west of the present town of Ilullisat, are responsible for the region's rich marine fauna that has made it possible for man to live in this harsh wilderness for thousands of years. This book describes the origin of Sermeq Kujalleq, the formation and drifting of the icebergs that fill Kangia, and the importance of the fjord to the hunting cultures of the past, and to commercial enterprise and tourism in modern Greenland.

In the summer of 2004, Ilulissat Icefjord was included on the World Heritage List. In 1972 UNESCO, the United Nations Education, Scientific and Cultural Organisation, established the World Heritage List to protect the world's most important cultural monuments and natural landscapes from damage or destruction, so that they can be preserved for the benefit and enjoyment of future generations.

A total of 788 sites in 134 countries are included on the list. Of these, 611 are sites of cultural importance and 154 are outstanding natural landscapes and 23 are mixed. Amongst the most famous are the Pyramids of Giza, the Great Wall of China and Yellowstone National Park in the USA.

It is the World Heritage Committee that decides which areas will be included on the World Heritage List, based on proposals put forward by individual countries. In 2002, Greenland's Home Rule government proposed Ilulissat Icefjord for inclusion, on the basis of the icefjord's scenic beauty and its unique importance for long-term glaciological studies.

Any locality included in the World Heritage List automatically receives enhanced recognition and status. As the name of the list implies, the site, wherever it is located, becomes not only the responsibility of the host country, but shares that responsibility with the entire population of the world on whose behalf it is preserved and protected. It is a condition of inclusion on the World Heritage List that the host country avoid activities that damage or diminish the value of the site, but in practice it is only poor countries that can expect to receive direct financial help for a site's preservation.

The stunning beauty of the Ilulissat region is due to the unique combination of the enormous production of icebergs from Sermeq Kujalleq at the head of Kangia and the presence of Isfjeldsbanken (the 'iceberg bank'), the threshold across the mouth of the fjord where the icebergs run aground. An unending drama of changing forms, colours and sounds is created as the icebergs melt, collapse and collide with each other. 'Ilulissat Icefjord' is the name used in the World Heritage List for the long iceberg-filled fjord known as Kangia (Ilulissat Isfjord), the surrounding land and a part of the Inland Ice of Greenland. At the head of Kangia lies Sermeq Kujalleq (Ilulissat Isstrøm), one of the most-productive glaciers draining the Inland Ice of Greenland.

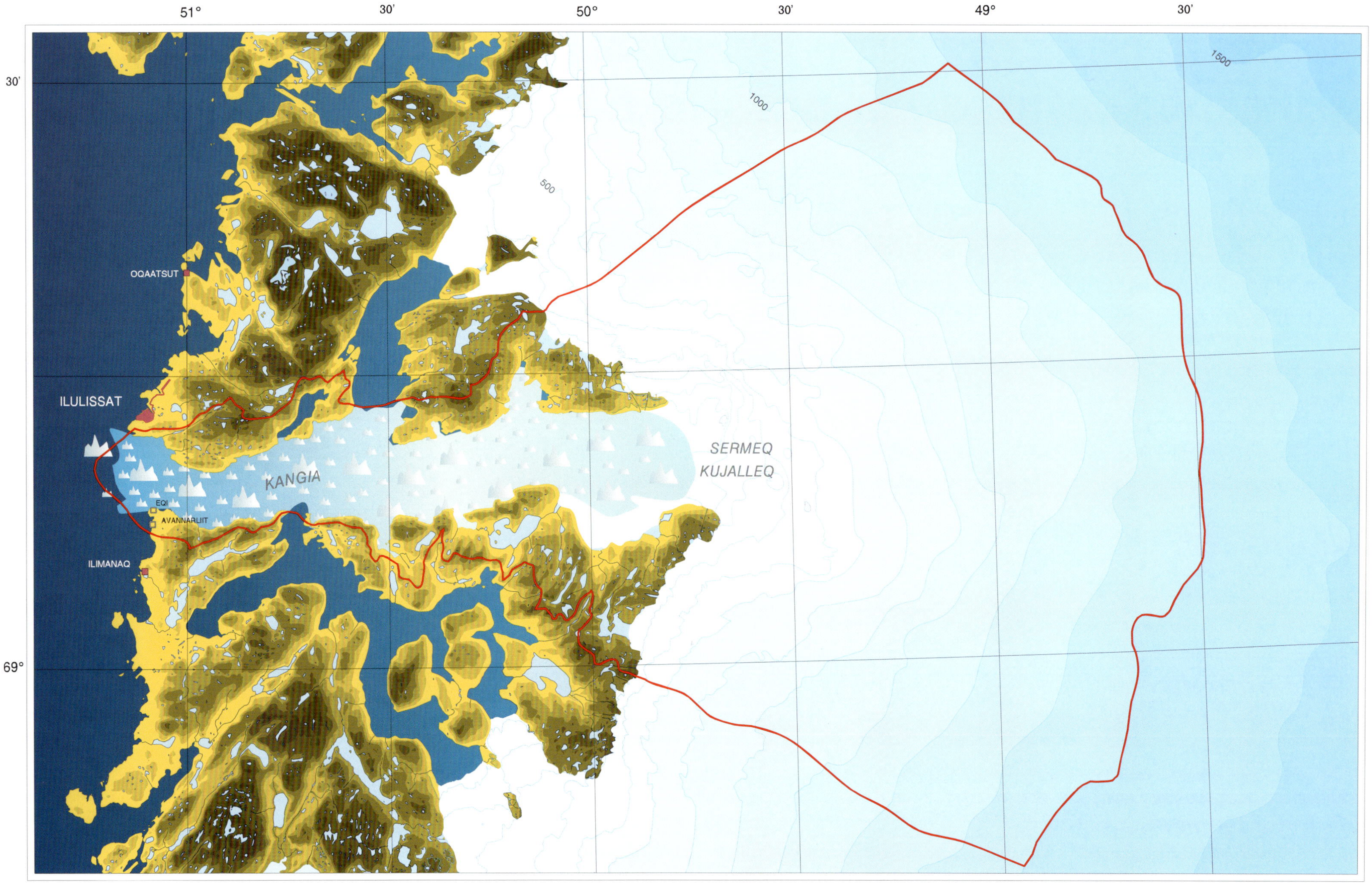

Map of the region around Kangia (Ilulissat Isfjord), with the limits of the Ilulissat Icefjord World Heritage Site shown in red.

Ilulissat - a modern town

The Greenlandic name for the town on the north side of the icefjord, Ilulissat, translates simply as 'icebergs': the community, the iceberg bank, the fjord and the fast moving glacier that gives birth to the icebergs are linked together. It is no coincidence that the town's coat of arms is a stylistic iceberg in silver-white and blue, adorned by two star-shaped ice crystals.

Ilulissat is the third-largest town in Greenland, with a population of more than 4000 people. Half of Greenland's fishing is centred here, and it is also home to the country's largest fish-processing factory. Today, Ilulissat is a living and modern town with crèches, kindergartens, schools including a music school, educational centres, leisure clubs, meeting places, a library, an old people's home, a 'socialpædagogisk seminarium', a hospital a power station and an airport.

The busy fishing harbour.

THE NEXT GENERATION OF SLEDGE DOGS SEEK SHELTER FROM THE WARM SUNSHINE

Large icebergs are visible from any point in Ilulissat.

Fishing is the main industry, and there is a large fleet of shrimp trawlers and smaller ships and boats for catching halibut. About one-quarter of the town's population is employed at the Royal Greenland fish factory.

Ilulissat is Greenland's leading tourist town, and is host to about half of all tourists that visit Greenland. The magnificent icebergs are the main attraction, but there are also many possibilities for excursions by foot or dog sledge, boat and helicopter. The surrounding countryside is unspoilt wilderness, where there are no plans for development.

4400 years of history

The Saqqaq Palaeo-Eskimo people settled near Ilulissat about 4400 years ago, but disappeared again after having lived around the fjord for 1500 years. For a short period West Greenland appears to have been uninhabited, until a new wave of Palaeo-Eskimo settlers, known as the Dorset culture, arrived about 2800 years ago; their dwellings sites are also found in the Ilulissat area. The Dorset culture disappeared from the region about 2000

Tourists on sledge tour in the area.

The Zion Church.

years ago, and the Disko Bugt region, including the Ilulissat area, was uninhabited for the next 1000 years.

The early Palaeo-Eskimo cultures used lightweight harpoons, sometimes with stone points, as well as lances and kni-

BOULDERS ON A FREE RIDE

ves. The Saqqaq people also used bows-and-arrows, and probably had kayaks. Sledges are first known from the Dorset culture, and were most probably drawn by the hunter himself or by a few dogs. Scrapers, drills, knives and needles of bone were used in their homes, a skin tent used all year around, with a fire-place in the centre. Driftwood from the Canadian and Siberian forests was an important resource, and a necessity for building tents, kayaks, sledges, bows and lances.

The Thule culture arrived in West Greenland around 1200 AD, and also settled in the Ilulissat area. Their winter houses were built from stone and turf, with a roof construction of driftwood or whale ribs. In the summer they lived in tents. A number of archaeological finds demonstrate that the Thule people had contact with the Norse, who lived in South-West Greenland from about 985 until around 1450 AD.

European explorers sailed into Greenland waters in the 16th century, followed by whalers in the 17th and 18th centuries. Both groups record contacts with the local Inuit population. The colonisation dates

Archaeological excavations at Qajaa. The layers in the midden date from the Saqqaq, Dorset and Thule cultures.

back to Hans Egede's voyage of 1721. Failing to find the lost Norse people he was seeking, Hans Egede founded a settlement near Nuuk, the first of many settlements established along the coast of West Greenland. The colony of Ilulissat / Jakobshavn was established in 1741. In 1979 Greenland achieved Home Rule, but it is still a part of the Kingdom of Denmark.

The Inland Ice and the icefjord

Greenland is the largest island in the world, with an area of 2.2 million square kilometres. The southernmost parts have a sub-arctic Atlantic climate, while the northern parts bordering the Arctic Ocean lie in the high arctic, and include the northernmost land areas in the world. The dominant feature of Greenland is the central ice sheet, the Inland Ice, which with an area of 1.7 million square kilometres covers 81% of Greenland and rises to an altitude above 3200 m.

The Inland Ice is the second largest body of ice in the world, and a relict of the Ice Ages. Deep drill cores through the Inland Ice have revealed detailed information of climate changes and the chemical variations of the atmosphere over the past approximately 250,000 years.

The present day ice-free strip of land bordering the Inland Ice was glaciated at the culmination of the last ice age about 21,000 years ago. At that time, the Inland Ice also covered large parts of the shelf areas off West Greenland. After the end of the last ice age, about 11,700 years ago, the Inland Ice slowly shrank to its present size, which has not changed very much over the past 9000 years. The moraines left behind by the retreating ice preserve evidence of advances and retreats of the Inland Ice margin, and are well seen in the Ilullisat area.

The Inland Ice occupies a bowl-shaped depression, the centre of which lies below sea level due to the load of the ice deforming the Earth's crust. The Inland Ice has a volume of 2.9 million cubic kilometres of ice. Each year snow equivalent to about 600 cubic kilometres of ice accumulates on the ice sheet, and there is an equivalent loss through the production of icebergs and melting. About half of the loss is due to the calving of icebergs from glaciers that reach sea level. The eastern and southern parts of the Inland Ice are limited by high mountain ranges, partially submerged in the ice. The central part of the Inland Ice therefore drains mainly towards the west, with several prominent outlet glaciers reaching Disko Bugt and Uummannaq Fjord. The area around Ilulissat represents the southern part of this drainage area and, as noted above, Sermeq Kujalleq is the most productive glaciers in Greenland.

The Inland Ice around Sermeq Kujalleq shows many features that are characteristic of the dynamics and mass balance in low arctic climatic conditions: areas with extensive melting and re-freezing, glaciers with widely different speeds and active calving. Other typical surface features are lakes and rivers on the ice sheet, and areas of crevasses.

Sermeq Kujalleq - the fast glacier

Sermeq Kujalleq is 3-6 km wide. The extremely high velocity of 35 metres a day, makes it one of the world's fastest glaciers. It is responsible for one-tenth of the total production of icebergs from the In-

MELTWATER RIVER NEAR THE MARGIN OF THE INLAND ICE

land Ice. The icebergs fill the fjord Kangia, and vary in size from small blocks to giants of 1.5 cubic kilometres of ice or more. Really large icebergs are rare, but they can always be seen at the mouth of the icefjord, where they run aground on the threshold of Isfjeldsbanken.

During the ice ages the forerunner of Sermeq Kujalleq filled the icefjord and eroded it to a depth of about 1000 m. Through the glacier's present activity, its meltwater transports nutrients to the fjord. The constant addition of nutrients forms the basis for the rich biological life in the fjord, which through the ages has created excellent hunting and fishing areas for the local people. The prehistoric hunting areas and settlements, remains of which are still found, were established and then abandoned as the location of the glacier front changed. In this way, the Ilulissat area illustrates the interplay between an ever-changing glacial environ-

The front of Sermeq Kujalleq seen from the north.

ARCTIC FOXES ARE QUITE COMMON IN THE AREA

ment and the development of human occupation.

Sermeq Kujalleq is one the world's best-studied glaciers and the studies have made crucial contributions to the understanding of fast moving glaciers, or ice streams. Tidal glaciers that calve in the sea and the resulting icebergs are natural phenomena, and the Ilulissat Icefjord World Heritage site preserves one of the best and most spectacular examples on Earth.

A rich fauna

Greenland halibut live in the entire fjord system. Off the mouth of Kangia halibut are plentiful and are subject to one of the most intense forms of fishing in the world. However, the interior ice-filled part of Kangia is usually difficult for extensive fishing, making it in effect an important 'protected' area for halibut, where they can become up to 25 years old.

Fin whales and minke whales are often seen at the mouth of Kangia in the summer. Humpback whales are uncommon, and blue whales and Greenland right whales are very rare visitors. Beluga and narwhale are found in Disko Bugt in the autumn and winter, and they occasionally swim into Kangia. Harp seals are common in open water in the summer, and the ringed seal is found in Kangia all year round.

Arctic plants close to the icefjord.

Polar bears are extremely rare visitors to the area.

The bird life is typical of fjords in central West Greenland. The rich marine life at the mouth of the icefjord attracts many foraging birds, especially fulmars and gulls. Land birds are less numerous than sea birds, but many of the common species in Greenland - snow bunting, Lapland bunting, redpoll, wheatear, Canada goose, raven - breed in the area. Greenland white-fronted goose, peregrine falcon, gyr falcon and ptarmigan are also found.

Iceberg aground on Isfjeldsbanken, the shallow threshold at the mouth of the icefjord. Former tilted water-level lines are clearly visible.

ICEBERGS AND BRASH ICE IN THE ICEFJORD

Among terrestrial mammals, the arctic fox is quite common. The arctic hare is mainly found in the higher-lying areas near the margin of the Inland Ice, and reindeer are found locally south of the icefjord.

In the spotlight of the world and science

The Inland Ice and the ice stream Sermeq Kujalleq that terminates in the interior of Kangia are celebrated natural phenomena that have attracted world-wide scientific interest since the first descriptions in the 18th century. The enormous ice-covered central region of Greenland has been a magnet for adventurers and explorers, whose attempts to 'conquer' the Inland Ice made headlines in the international press of the time. Fridtjof Nansen's crossing of the Inland Ice in 1888 was a sensational achievement, and the first of many. Today sporting expeditions regularly cross the Inland Ice, and continue to provide good stories for the media.

Hinrich Rink published the earliest scientific descriptions of the Inland Ice in the middle of the 19th century. His articles aroused interest in scientific circles, because they supported the argument that ice sheets had formerly covered large areas in northern Europe and North America. Today, the ice sheets in Greenland and in Antarctica play a central role in studies of earlier climate changes and their global impacts. The Inland Ice of Greenland is the only body of ice in the Northern Hemisphere to preserve a continuous registration of climatic changes for the past 250,000 years.

Fridtjof Nansen's 1888 expedition was the first to cross the Inland Ice.

The relationship between the accumulation of snow in the high interior parts of the Inland Ice and the loss through melting and calf-ice production along its marginal parts, as well as the dynamics of the glaciers, plays a central role in studies of how large ice masses react to local and global climate changes. The Greenland Inland Ice is more sensitive to climate change than the Antarctic ice sheet, and is therefore better suited for studies of the connection between ice sheet dynamics and climate change. From north to south the Inland Ice is characterised by widely different climatic conditions, that facilitate climate-related studies of the ice sheet. These studies

Watercolour: A. Kornerup, 1878

Many expeditions explored the Inland Ice in the 19th century. Their equipment was primitive, and travel on the ice sheet was difficult and dangerous.

WHEN ICEBERGS MELT, SPECTACULAR FORMS ARE CREATED

make a crucial contribution to our understanding of how the large ice sheets in Europe and North America disappeared at the end of the last ice age. At the same time, the studies can provide information about how ice sheets will react to future changes of the Earth's climate.

The ice sheets of Greenland and the Antarctica are the only remnants of the widespread continental ice sheets of the last ice age. The present Greenland Inland Ice covers an area of 1.7 million km^2 and is up to 3200 m thick. The oldest ice is found at the bottom of the central part of the ice sheet and is about 250,000 years old. The Antarctic ice sheet contains almost ten times as much ice as Greenland, covers an area of 13.5 million km^2, while the oldest ice is believed to be about a million years old.

All the other glaciers in the world are younger and smaller. Even the largest local ice caps are dwarfed by the Greenland ice sheet. Because of the old age of the ice and their great size, the ice sheets in Greenland and the Antarctic preserve information about glaciological and glacio-geological processes that are not found in the remainder of the world's present glaciers. The studies of these special characteristics are of vital importance for the understanding of ice age processes, climatic history and the relationships between global climatic changes and the major ice sheets.

There is an important difference between the Greenland and Antarctica ice sheets in respect of their interplay with climate. This is because the whole of the Antarctic ice sheet lies in the high arctic climate zone, where there is little melting even in summer. By contrast, the Greenland Inland Ice extends from the high arctic in the north to the sub-arctic and temperate climate zones in the south. For this reason, the Greenland Inland Ice reacts dynamically to a broader range of climatic conditions.

The major ice streams

The ice streams of the Inland Ice are smaller than the gigantic ice streams of the Antarctic. In Greenland, only a few outlet glaciers are true ice streams, and Sermeq Kujalleq is by far the most important of them. Sermeq Kujalleq has a pronounced glacier structure with a heavily crevassed surface that is due to the permanent high speed of the ice stream of about 7 km a year, almost one metre per hour. Recently, the velocity of Sermeq Kujalleq has almost doubled.

In the Northern Hemisphere there are also fast flowing ice streams in Alaska. The Columbia Glacier reaches speeds of up to 11 km per year, and the LeConte Glacier about 10 km a year. Both glaciers, as well as Sermeq Kujalleq, have been subject to recent rapid retreat.

The annual production of calf ice from Sermeq Kujalleq is about 35 cubic kilometres, equal to a tenth of the total calf ice production of the Greenland Inland Ice. Its drainage area is 6-7 % of the total area of the Inland Ice. No other glacier in Greenland plays so important a role in the ice-transport from the interior of the Inland Ice to the coast.

Sermeq Kujalleq is the fastest permanently flowing ice stream in Greenland. The only glaciers that occasionally achieve higher velocities are those that show periodic surging, i.e. glaciers that make rapid advances for short periods. Surging glaciers usually only have sufficient ice in their drainage areas to feed short advances at high speed. Several of the major glaciers draining from the Inland Ice into Disko Bugt and Uummannaq Fjord are also considered to be permanent high-speed glaciers (ice streams). A few of the large outlet glaciers of the Inland Ice in North and East Greenland move at varying velocities, classified as pulsating movements.

Although smaller in size than the gigantic ice streams of the Antarctic, Sermeq Kujalleq is still a major ice stream with respect to the quantity of ice discharged. The Antarctica ice streams are nearly all an order of magnitude slower than Ser-

AREAS AND LOCATIONS OF SOME OF THE LARGER GLACIERS AND ICE FIELDS

Name	Location	Co-ordinates		Surface area (km^2)
ICE FIELD RANGES	Alaska and Yukon	61°N	141°W	c. 60,000
U.S. RANGE ICEFIELD	Ellesmere Island, Canada	82°N	76°W	c. 28,210
AGASSIZ ICE CAP	Ellesmere Island, Canada	81°N	73°W	c. 20,400
NOVAYA ZEMLJA	Barents Sea, Russia	76°N	62° E	c. 19,000
VESTSPITSBERGEN	Svalbard, Norway	79°N	18° E	c. 15,300
PATAGONIAN ICEFIELD	Patagonia, Chile	50° S	74°W	c. 13,000
VATNAJÖKULL	Iceland	64°N	17°W	c. 8,400
SIACHEN GLACIER	Pakistan/India	36°N	77° E	c. 1,000

meq Kujalleq, and therefore part with their ice at a lower rate.

Ice calving

An important difference between Sermeq Kujalleq and Antarctic glaciers is the manner in which the ice calves. In the Antarctic, ice is often discharged in major calving events, when the glacier front retreats several kilometres over a short period. Until the sudden 2002 collapse, calving of icebergs at Sermeq Kujalleq has been an almost continuous process, with large tabular icebergs only released in the summer season.

In the Antarctic, floating ice shelves are common, and a considerable part of the Antarctic ice disappears through melting at the bottom of these floating ice shelves. The bottom of a floating ice shelf can melt at up to 5-10 metres a year.

Observations at Sermeq Kujalleq extend back over the past 150 years, and provide evidence of its reaction to climatic variations. No other ice stream in the world has an equally long series of observations of its frontal positions. Although the Antarctic ice streams have been studied intensively during the past 50 years, Sermeq Kujalleq continues to play an important role in studies of the dynamics of rapidly moving glaciers, and the effects of climatic change on the drainage of ice sheets.

Map showing the location of World Heritage sites with glaciological features and glaciers mentioned in the tables.

LARGEST ANTARCTIC ICE STREAMS

Name	Discharge	Velocity at grounding line
THWAITES GLACIER	51 km³/year	3.00 km/year
PINE ISLAND GLACIER	68 km³/year	1.50 km/year
ICE STREAM B	17 km³/year	0.80 km/year
FOUNDATION/MÖLLER GLACIER	51+25 km³/year	0.55 km/year
TOTTEN GLACIER	72 km³/year	
LAMBERT GLACIER	58 km³/year	
BYRD GLACIER	55 km³/year	0.80 km/year
ICE STREAM E	26 km³/year	0.55 km/year

TURQUOISE COLOURED ICE SCULPTURES

Protection and management

The World Heritage site is protected, as laid down by Greenland Home Rule order no. 7 of 25th March 2003. In addition, certain activities have been regulated through hunting and fishing legislation, environmental legislation and the like.

The national protection of the Ilulissat Isfjord region aims at:

- Ensuring the integrity of the area for coming generations; changes to or construction of buildings that might impair the scenic character of the region are not allowed
- Regulating visits to and transport within the protected area by vehicles, helicopters and dog sledges
- Regulating tourism
- Protecting cultural heritage sites

In addition, Ilulissat town council has adopted an administrative plan for the area to regulate local activities and tourism by:

- Managing the areas where tourist huts are located
- Controlling the treatment of house hold rubbish
- Marking paths near urban areas and in the vicinity of tourist huts
- Designating and managing permanent camping areas
- Managing fishing and hunting cabins

The government of Greenland - 'Landsstyret' - has the overall responsibility for protecting Greenland's nature and environment. The Directorate of the Environment and Nature and local authorities administer this responsibility. Ilulissat municipality has a technical services office that has the day-to-day responsibility for ensuring that the provisions of the management plan for the World Heritage site of Ilulissat Icefjord are observed. In addition, the Home Rule Office has recruited hunting and fishing gamekeepers, whose task is to ensure that the hunting and fishing regulations are observed. The National Museum and Archives of Greenland is the authority responsible for protecting and preserving cultural sites.

Giant ice wall with large crevasses

Science born at the Inland Ice and the icefjord

MEETING BETWEEN INUIT AND EUROPEANS

The Inland Ice has unique historical importance for the foundation of glaciology - the science of glaciers - and the gigantic ice sheet of Greenland still plays a central role in modern glaciology and climate research. Published information about the Inland Ice and the present Ilulissat Icefjord World Heritage site goes back more than 250 years.

Hinrich Rink (1819-1893) was a pioneer in studies of the Inland Ice, and the first to realise its enormous size.

The first descriptions of the Inland Ice date from the early Middle Ages. 'Kongens Spejl' (the King's mirror), a Norwegian document written about 1260, provides a vivid first-hand description of the widespread extent of Greenland's ice sheet:
"But as you asked whether the land is free from ice or not, or if it is covered with ice like the sea, then you should know that there is a small part of the land which is free from ice, but the rest is covered by it, and people do not know whether the land is large or small, because all fell areas and all valleys are hidden by the ice, so it is not possible to find an opening". It continues *"People have often tried to go to different places on the land on the fells that are highest, to look about and try to find any part of the land that was free from ice and habitable, but nowhere has it been possible to discover such areas apart from those that are already inhabited, which stretch only for short areas along the coast".*

The period of discovery: 1700-1845

The initial colonisation of Greenland in 1721, and the establishment of Danish settlements (Ilulissat / Jakobshavn was founded in 1741), resulted in many descriptions of the country, the nature and the Inuit people.

Missionaries and traders who travelled along Greenland's west coast wrote about the area around Ilulissat and the Inland Ice. Observations from the start of the 18th century indicate that the icefjord Kangia was then occasionally ice-free, whereas the margin of the glacier advanced in the 18th century and the start of the 19th century.

In 1770, the missionary David Cranz was one of the first to try to understand the origins of the Inland Ice and the icebergs, and he compared his observations with

European and American glaciers. This provided the foundation for more detailed surveys of the Greenland ice sheet during the following century, when it was realised that glaciers could also yield information about earlier changes between warm and cool periods, a sort of 'climatoscope'.

Observations and theories: 1845-1950

One person dominated the early years of true glacier research in Greenland for decades: Hinrich Rink. Rink was originally educated in chemistry and physics, but he carried out geological and glaciological studies in West Greenland between 1848 and 1852 during travels along the coast from Upernavik in the north to Qaqortoq / Julianehåb in the south.

The margin of the Inland Ice at Paakitsoq north of Kangia, drawn in 1850 by Hinrich Rink, one of the pioneers of glaciology. The ice margin was then advancing, and reached a maximum about 1880. Since then a gradual retreat has taken place, as is seen from the photographs from 1961 and 1987.

Drawing: H. Rink

Photo: Anker Weidick

Photo: Henrik Højmark Thomsen

DIRTY AND CLEAN ICE SIDE BY SIDE

Hinrich Rink's objective was to make a map of the country based on his own and earlier surveys, and supplemented by map sketches made by local hunters. He concentrated his efforts around the central part of West Greenland, an area that was then largely unknown, and he succeeded in surveying large parts of the fjords and their glaciers. Rink spent the winter of 1850/51 in Ilulissat and sailed to the margin of the Inland Ice at Paakitsoq, north of the town, in October 1850.

He travelled by sledge to the southern part of Sermeq Kujalleq in April 1851, and here he determined the position of the front of the glacier. Rink's surveys were the first in a series of observations of changes of the ice margin that have now lasted more than 150 years. Meteorological observations at Ilulissat were begun in 1873, and together with measurements of the movements of Sermeq Kujalleq from 1875, provide unique documentation of glacier frontal variations in arctic areas.

Through his work, Hinrich Rink became the first scientist to realise the vast extent and special form of 'the ice plain' covering the entire interior of Greenland, and he called this ice cover Indlandsisen, ('Inland Ice') after a suggestion by the Danish scientist Japetus Steenstrup. The Inland Ice was completely different from the local glaciers of northern Europe,

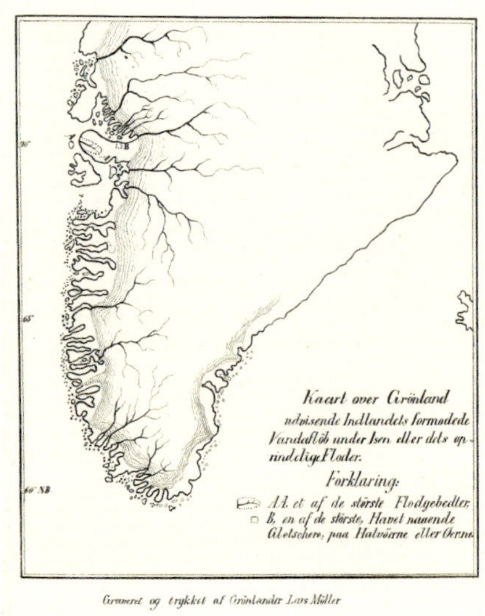

Map illustrating Rink's 1862 theory of the drainage of the interior of Greenland by rivers. As the map shows, East Greenland was then not well known, and the depiction of the area north of Ilulissat in West Greenland is also imprecise.

Rink's 1857 map showing the extent of the Inland Ice, and the five large glaciers (ice streams) from the Inland Ice that drain into Disko Bugt and Uummannaq Fjord. Less important glaciers in southern Greenland are also shown.

Source: H. Rink, 1857 and 1862

well known from many descriptions. At about the same time a new idea was developing in Europe - that ice had once covered large parts of northern Europe in the past. Rink's sensational news that there was an enormous ice sheet in Greenland was therefore received with particular interest in scientific circles, because it could substantiate the argument that northern Europe had been covered by ice during one or more ice ages. Rink's descriptions also provided the impetus for glaciological expeditions to Greenland and the Ilulissat area in the following decades.

Hinrich Rink also tried to understand the mechanism behind the production of calf ice. His main theory was that the ice streams flowing from Greenland's interior are fed by the accumulation of snow. The snow, in the form of ice, is then transported out to the coast by the ice streams - in the same way that rain over large land areas is carried to the sea by large rivers.

Rink estimated the production of calf ice on the basis of the size of the glaciers and the quantity of ice in the fjords, and he recognised five icestreams of 'the first order' - Jakobshavn, Tossukatek, Den større Kariak, Den større Kangerdlursoak and Upernavik. He thus realised that there was large calf ice production in five fjords, although he did not discover the unique status of Sermeq Kujalleq as the main outlet of the Inland Ice. That was first realised in the second half of the 19th century. However, Rink recognised the extraordinary size of the Inland Ice, which is indicated by his calculation of the catchment area of about 50,000 square kilometres or more that he estimated was necessary to feed each of the five glaciers.

BOX 2: 150 YEARS OF EXPLORATION

1850/51: H. Rink was the first to determine the frontal position of Sermeq Kujalleq.

1870: N.A.E. Nordenskiöld described the front of Sermeq Kujalleq, but was unable to determine the boundary between the glacier front and the calf ice in the fjord. His visit to the glacier took place during one of the earliest attempts to reach the unknown interior parts of Greenland. During a reconnaissance from the bottom of Arfersiorfik fjord, 90 km south of Ilulissat, he traversed 56 km across the Inland Ice reaching an altitude of 680 metres above sea level. A later larger expedition led by Nordenskiöld, from the same starting point, reached 350 km into the Inland Ice and an altitude of 1947 metres. The main purpose was to investigate the ice sheet, and study the landscape at the ice margin.

Nils Adolf Erik Nordenskiöld

1875: A. Helland determined the location of the Sermeq Kujalleg glacier front, and made the first measurements of the velocity in July 1875.

1879: R.R.J. Hammer surveyed the whole of the fjord system, including the glaciers at the bottom of the Sikuiuitsoq and Tasiusaq fjords. He also determined the position of the Sermeq Kujalleq front in September.

1880: R.R.J. Hammer repeated his visit to the glacier in March and August 1880. Hammer recorded a winter advance of 1 km for Sermeq Kujalleq, and a subsequent summer retreat of 2 km. He also measured the movement velocity from the southern edge of the glacier front to its central part, and estimated that a movement of up to 16 metres in 24 hours was possible. Hammer reported that the movement was not uniform, and that there appeared to be a connection between air temperature and glacier velocity. The position of the front was very variable, and Hammer concluded that large icebergs were released through a break-up of the ice front due to the buoyancy of the floating part of the glacier.

1886: R.E. Peary attempted to reach the interior of the Inland Ice from a starting point at Paakitsoq, 40 km north of Ilulissat. Accompanied by Christian Majgaard he reached a point 185 km into the Inland Ice and an altitude of almost 2300 metres above sea level. A description of the landscape near the ice margin was published.

Robert Edwin Peary

1888: S. Hansen made a sketch map of the icefjord for the Royal Danish Sea Chart Archive. According to Engell, a photograph made during this visit showed that the position of the glacier front was further to the east than shown on the map of this year.

1893: E. von Drygalski visited the area to the south of the Sermeq Kujalleq front in February. The position of the front was drawn on a map, although it was difficult to determine precisely.

1902: M.C. Engell measured the frontal position of Sermeq Kujalleq in July, and showed that the process of retreat since Rink's observations in 1851 had continued. Velocity measurements were similar to those previously determined. Engell also described the whole fjord region, including Sikuiuitsoq and Tasiusaq.

1903: M.C. Engell again visited the glacier in July, but could only determine the frontal position with some uncertainty; it appeared to be about 350 metres further west than in 1902. He reported the calving of a large iceberg.

1904: M.C. Engell made his third visit to the glacier front in the summer of 1904. The position of the front was not determined.

1913: J.P. Koch and A. Wegener. After their east to west crossing of the Inland Ice in 1912-1913, the two scientists visited Sermeq Kujalleq in August 1913. Their determination of the position of the glacier front indicated a considerable retreat since 1902

1929: J. Georgi and E. Sorge. During the preparations for Alfred Wegener's Expedition to the Inland Ice in 1930-1931, two of its members determined the position and movement of the ice front during a visit in autumn 1929.

1931/32: The position of the glacier front was depicted on the first map sheet (1:250,000) of the Jakobshavn area, published by the Geodetic Institute, Copenhagen.

After 1932: During and after the Second World War there were rapid developments in aerial photography techniques. Aerial photographs of Kangia and Sermeq Kujalleq in the archives of the National Survey and Cadastre in Copenhagen date from the years 1942, 1946, 1948, 1953, 1957, 1958, 1959 and 1985. In more recent years satellite imagery allows continuous monitoring of frontal positions.

IN THE SUMMER, THE SUN SHINES ALL NIGHT

Interest in Greenland's glaciology grew in the second half of the 19th century, with an increasing number of researchers visiting Kangia and Sermeq Kujalleq; nearly all described the glacier front and its surroundings. Among the early visitors were some of the leading scientists and explorers of the time, including Erich von Drygalski, Robert E. Peary, Adolf Erik Nordenskiöld and Alfred Wegener.

Modern glaciology from 1950

During the Second World War map-making on the basis of aerial photographs was developed to a high standard, so that even remote areas such as Greenland could be mapped in detail from aerial photographs. At the same time, all-terrain vehicles were developed, which eased access to, and transport on, the inner crevasse-free areas of the Inland Ice.

The modern era started with the activities of Expéditions Polaires Francaises in the years from 1948 to 1953. The expedition's work was focused along an East - West profile from coast to coast over central Greenland, with the purpose of measuring a profile of the bottom and surface of the Inland Ice and determining the mass balance along the route. The west end of the profile was 70 km north of Ilullisat, on the south side of Eqip Ser-

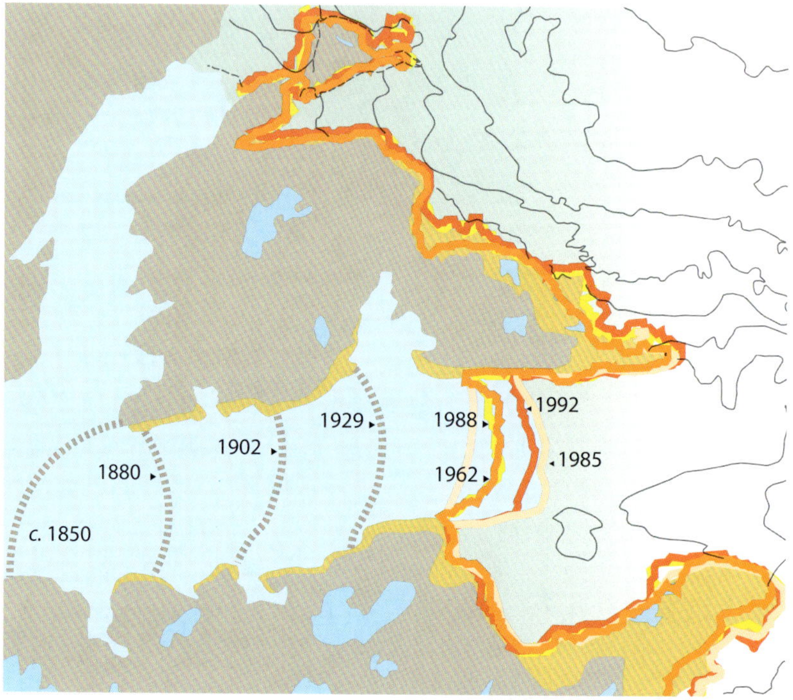

Sermeq Kujalleq gradually retreated between 1850 and 1992. The glacier front was more or less stable between 1950 and 2002, with seasonal fluctuations, retreat in summer and advance in winter. The 1962 and 1988 positions shown are respectively winter and spring locations of the front. The 1985 and 1992 positions show the location of the front in respectively summer and early autumn. The 1992 determination, from satellite data, also depicts the ice margin to the north and south of Sermeq Kujalleq. *See also page 53.*

mia glacier, and this area was the subject of detailed mapping. However, there was little direct contribution to knowledge of the Ilullissat region.

Expédition Glaciologique Internationale au Groenlande in 1957-1960 was a collaboration between Austria, Denmark, France, Germany and Switzerland. Using aerial photographs, the expedition determined the rate of movement of all the outlet glaciers draining into Disko Bugt and Uummannaq Fjord. At the same time the production of calf ice was assessed, and the studies confirmed Sermeq Kujalleq's unique role with regard to the production of calf ice.

After 1960, an important project was undertaken to determine the thickness of the Inland Ice and map the altitude of the ice surface and the topography of the landscape under the ice using airborne radar. More than 60,000 kilometres of profiles were collected over six seasons between 1968 and l976 - the first overall survey of the landscape under the Inland Ice. The survey was a breakthrough in the understanding of the dynamics and geomorphology of glaciers and ice sheets, and in climate research.

However, there was still a lack of detailed knowledge about the continuation of Kangia and the channel occupied by Sermeq Kujalleq beneath the margin of the Inland Ice. Some of this information has since been collected during investigations of the potential for using melt water for hydropower production. For this purpose a helicopter-borne radar technology was developed by the Technical University of Denmark, and modified by the Geological Survey of Denmark and Greenland. The most recent surveys of the thickness of the ice sheet around Sermeq Kujalleq were published in 2001.

CALF ICE PRODUCTION ON THE GREENLAND WEST COAST

Glacier	Position	Production	Velocity
Sermeq Kujalleq in Kangia	69° 11´N	c. 35 km³/year	c. 6.9 km/year
Sermeq Kujalleq in Torsukattak	70° 00´N	8-10 km³/year	2.6-3.5 km/year
Store Gletscher	70° 20´N	13-18 km³/year	4.2-4.9 km/year
Rink Isbræ	71° 45´N	11-17 km³/year	3.7-4.5 km/year
Gade Gletscher	76° 20´N	c. 10 km³/year	--

Data from the 1960s, localisation page 51

Modern studies of the marginal areas of the Inland Ice are, to a large degree, based on data from satellites. Satellite monitoring of frontal changes has replaced aerial photography to some extent as it makes it possible to carry out repeated observations within a short time frame. The record of Sermeq Kujalleq's development since 1932, established from aerial photographs, has been one of steady retreat until the 1950s. Then followed a period of relative stability until 2002, when satellite monitoring has documented a rapid retreat.

Glacier surges

During the second half of the 20th century increasing interest was shown by glaciologists in the phenomenon of glacier surging, during which glaciers make short term advances at relatively high velocities. This phenomenon has been described for local glaciers at different places around the world. Surges have velocities comparable to the high velocities of the so-called ice streams of Antarctica and Greenland, the major drainage outlets of the ice sheets for which the high velocities are permanently maintained due to the large catchment areas.

CASTLES OF ICE WITH TOWERS AND SPIRES

The interest in understanding surging glaciers is related to the development of the idea that the ice sheets of the ice age may have collapsed suddenly and catastrophically. Similar considerations are relevant today with respect to the possible response of the present ice sheets to global warming as a result of the increased greenhouse effect caused by mankind. Interest in the dynamics of glaciers has therefore increased. Sermeq Kujalleq is an obvious choice for studies, partly because the region is covered by a long series of climate observations from the weather station at Ilulissat, but also because historical and scientific observations extend back for 150 years. In addition, due to studies relating to hydropower, a continuous series of temperature, precipitation and radiation measurements were made at the margin of the Inland Ice in the 1980s, to determine the quantity of melt water produced from the ablation zone.

Information about the most recent changes in extent and movement of glaciers is based on surveys from satellites, which can also determine the topography of the ice and the surface velocity over large areas. The dramatic retreat of the floating segment of Sermeq Kujalleq in 2002 will undoubtedly lead to more intensive studies in the near future.

The 40-90 m high front of Sermeq Kujalleq.

The Inland Ice and glaciology

MANY COLOURS AND SHAPES ARE FORMED DURING THE MELTING

The Greenland Inland Ice is the only major remnant of the gigantic ice sheets that covered large parts of the continents and the continental shelves of the Northern Hemisphere during the ice ages. The Inland Ice, together with the far larger Antarctic ice sheet, make up 99% of the volume and 97% of the area of the world's glacier cover

Greenland's Inland Ice has a surface area of 1.7 million square kilometres and is up to 3.2 km thick. The accumulation of snow equal to about 600 cubic kilometres of ice a year. Recently, the loss through calving and melting in the marginal zone has surpassed the accumulation, and the Inland Ice is retreating.

The Inland Ice occupies a bowl-shaped depression in the Earth's crust, formed by the weight of the ice cover. If the Inland Ice disappeared, the land surface would rise, and form a landscape similar to that of Greenland prior to the formation of the ice sheet. Large rivers would drain the inner parts of Greenland, with some of the largest rivers draining westwards towards Disko Bugt and across the continental shelf off West Greenland.

After the onset of glaciation in Greenland, the river valleys were further eroded at the margin of the ice sheet; the old river valleys can still be traced beneath the margin of the Inland Ice. In some areas, the river valleys were overdeepened by the ice and functioned as drainage routes for the outlet glaciers during various stages of the development of the ice sheet. Calf ice production then, as today, was concentrated to the large glaciers occupying these former river valleys. The valleys are now found as troughs on the continental shelf, as fjords in the ice-free land areas and as deep depressions

About 21,000 years ago, during the last ice age, the large ice sheets of the Northern Hemisphere covered an area of 30 million km², including much of North America. Ilulissat was centrally placed in the glaciated regions of the Northern Hemisphere.

Left:

The Inland Ice today. The dark blue region shows the drainage area of Sermeq Kujalleq.

Centre:

Topography of the bedrock below the Inland Ice.

Right:

Greenland without the present ice sheet, and with altitudes corrected for the present load of the ice sheet. Large parts of central Greenland are drained by rivers flowing into Disko Bugt in West Greenland. The continuation of the old drainage pattern can be seen as troughs in the offshore areas.

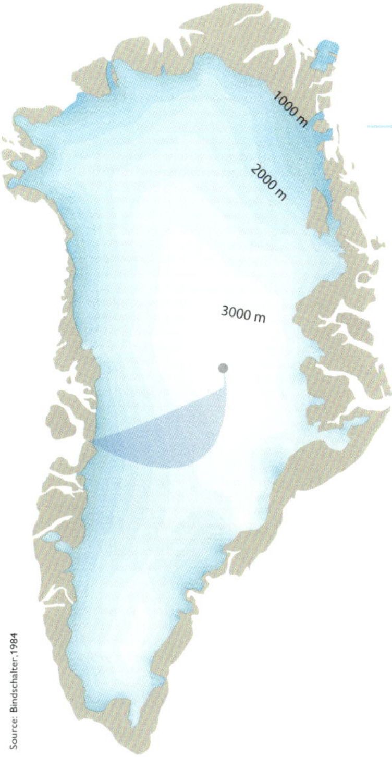

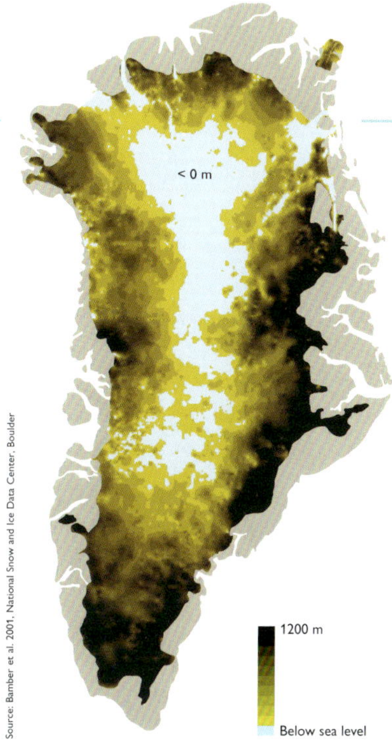

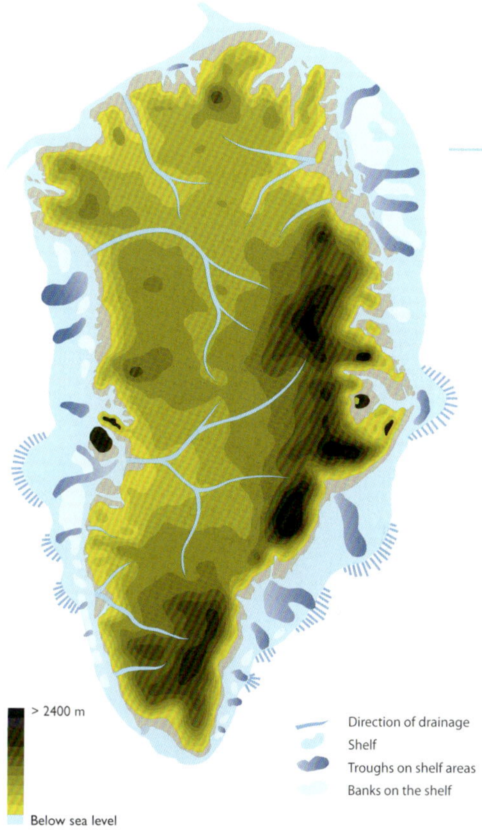

beneath the major ice streams in the marginal parts of the Inland Ice.

Although calving glaciers are widespread along the 6000-kilometre long perimeter of the Greenland ice sheet, calf ice production from the Inland Ice is concentrated in a few, large outlet glaciers. Many of the most important outlet glaciers are found along the coast of West Greenland, and 100 cubic kilometres of calf ice stem from the five largest outlet glaciers. The rest of the calf ice originates from numerous smaller glaciers.

Earlier glaciations

The transition from the ice-free landscape of pre-glacial times to Greenland's present glacial landscape is not known in detail. Data from sediment cores from the sea-bed indicate that there was a pronounced global cooling about 38-24 million years ago, which led to the formation of an ice sheet in the eastern Antarctic. After a relatively warm period, which culminated 15 million years ago, the climate started to get colder again about 10 million years ago. Sediments deposited in the North Atlantic about 7 million years ago contain material that was probably transported by icebergs, and this is taken as the first indication of widespread glaciation in southern Greenland. Sediments with ice-rafted material deposited about 2.5 million years ago, are considered to be indicative of even more widespread glaciation in Greenland.

GLACIER ICE WITH A HIGH CONTENT OF SAND, GRAVEL AND STONES

Ice cores drilled through the Inland Ice preserve an archive of past climate variations. The Nord-GRIP boring in northern Greenland reached the bottom of the ice sheet, and yielded a 3085 m long ice core that goes back 120,000 years in time.

The Inland Ice preserves a unique archive for the climatic evolution and atmospheric chemistry of the past 250,000 years. Snow that falls on the interior part of the ice sheet in one year is buried by the snows of successive years, and is gradually converted into ice by the pressure of the overlying snow. In the course of thousands of years the layers of ice sink deeper and deeper into the ice sheet.

The remarkable thing is that each year's layer of ice 'remembers' the temperature from the time that the snow fell. The reason for this is that the relationship between two oxygen isotopes in the snow reflects the air temperature during the snowfall. The isotopes are stable and are preserved in the ice, and it is therefore possible to follow the variations in climate backwards in time by analysing progressively older and older ice layers.

To date, five deep holes have been drilled in the Inland Ice. The ice cores recovered are kept frozen, and flown back to laboratories where the annual ice layers are analysed and dated. In this way information about the climate in Greenland during the past 250,000 years has been documented, a continuous record that reaches back through the last ice age.

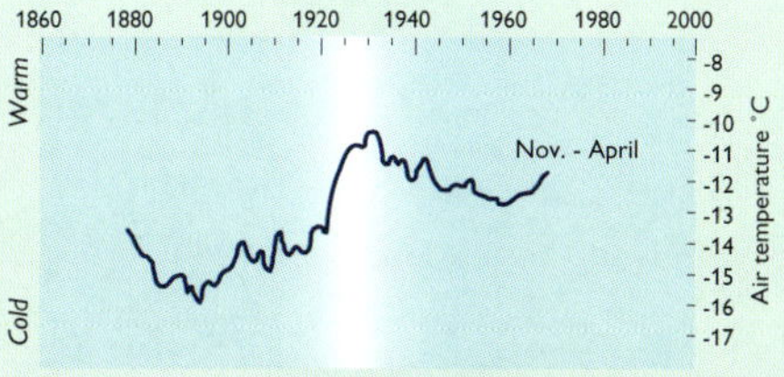

The development in winter temperatures at Ilulissat during the time period from 1878 to 1971. A marked rise in temperatures is seen between 1920 and 1925.

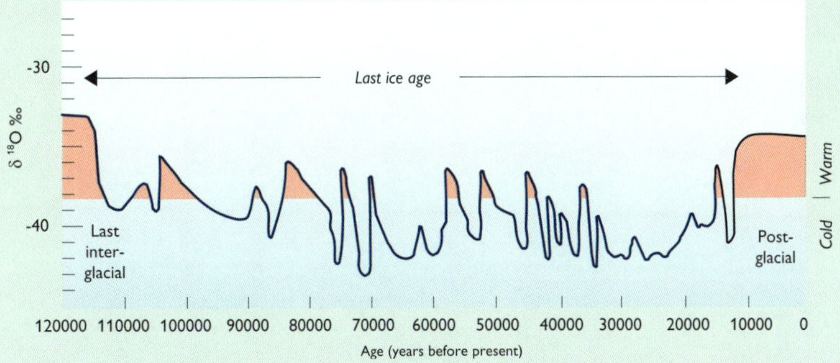

Changes in the Greenland climate over the past 120,000 years.

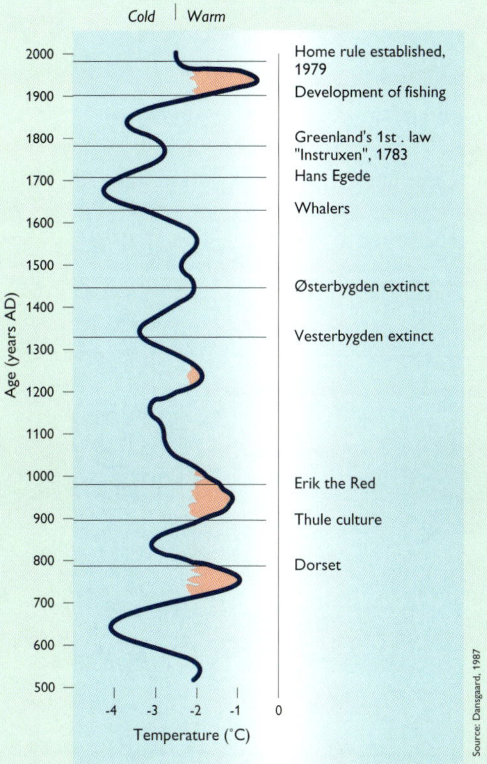

Climate curve for the past 1500 years, and important events in the history of Greenland.

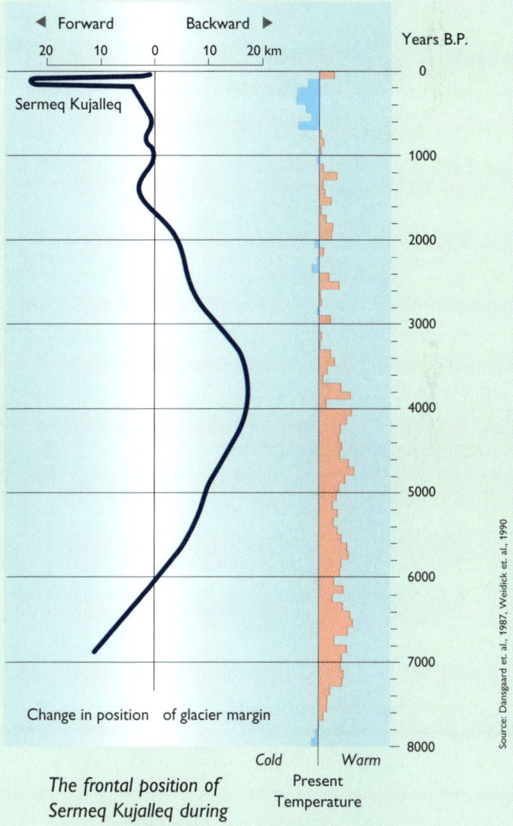

The frontal position of Sermeq Kujalleq during the past 8000 years reflects the climatic variations.

BOX 3b: THE UNSTABLE ICE-AGE

The drill cores show that during the last ice age the climate was extremely unstable. Periods of intense cold were interrupted many times by milder periods that sometimes lasted several thousand years. Typically, the climate suddenly became 10-20 degrees warmer in the course of a few decades, and then became slowly colder again in the course of a few millennia. It is thought that the marked oscillations in climate reflect changes in global ocean currents, especially the North Atlantic Current.

ERUPTION OF THE VOLCANO PINATUBO IN THE PHILIPPINES ON THE 12TH JUNE 1991. THE ASH COLUMN IS 23 KILOMETRES HIGH. 30 MILLION TONS OF ASH AND GAS WAS EJECTED, OF WHICH SULPHUR DIOXIDE MADE UP 20 MILLION TONS.

The last ice age ended abruptly about 11,700 years ago, and the climate in the present warm interglacial period has been relatively stable. The warm interglacial culminated from 8000 to 4000 years ago, and was followed by falling temperatures with two particularly cold periods about 1600 and 1850, the so called 'Little Ice Age'; average temperatures were then 1°C colder than today.

It is not only information about past temperatures that is preserved in the Inland Ice. Small air bubbles caught in the ice as the snow was gradually buried preserve the former composition of the atmosphere. It is thus possible to determine the content of the so called 'greenhouse' gases in the atmosphere as far back in time as the ice core extends.

Major volcanic eruptions can be traced in the Inland Ice as layers of ice with relatively high acidity, since large amounts of sulphuric acid are formed during volcanic eruptions. There may also be small particles of volcanic ash in the ice. It is known that large, or prolonged, volcanic eruptions have a cooling effect on global climate, since the ash particles and the small sulphuric acid drops in the atmosphere formed during a volcanic eruption reflect sunlight and thus energy from the sun back into space. The concentration of dust particles in the ice reflects the number and size of past storms. Dust from deserts and steppes is transported widely in the atmosphere, and it is known that dust that reaches Greenland today can come from central Asia. Dust content in the ice from the last ice age are substantially higher than in ice from the current interglacial period.

***A:** Cross-section of the Inland Ice, showing flow lines from the ice divide in the centre towards the ice margins; the flow in the ice sheet is determined by the slope of the surface. Ice cores have been drilled in the middle of the Inland Ice, where the annual layers are horizontal and easier to differentiate.*

***B:** The GRIP ice core was drilled at Summit, the highest point of the Inland Ice, and contains information on climate variations throughout the past 250,000 years. The curve shows the oxygen isotope ratios for the lowest 1500 m of the ice sheet. Low isotope ratios indicate cold periods.*

***C:** The movement of the ice transports the oldest ice to the lowest areas of the ice sheet. At the ice margin the age of the ice becomes progressively younger inwards towards the equilibrium line at about 1200 m above sea level. The curve shows an oxygen isotope record from the margin of the Inland Ice at Paakitsoq, 27 km North-west of the front of Sermeq Kujalleq, for the past 150,000 years. The Younger Dryas cold period marked the last cold phase prior to the start of the present interglacial period.*

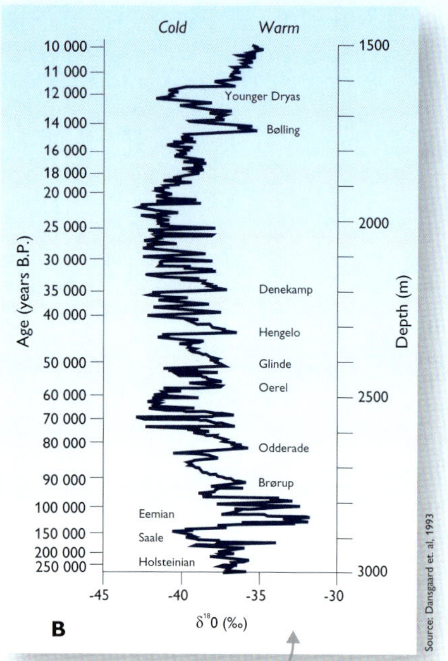

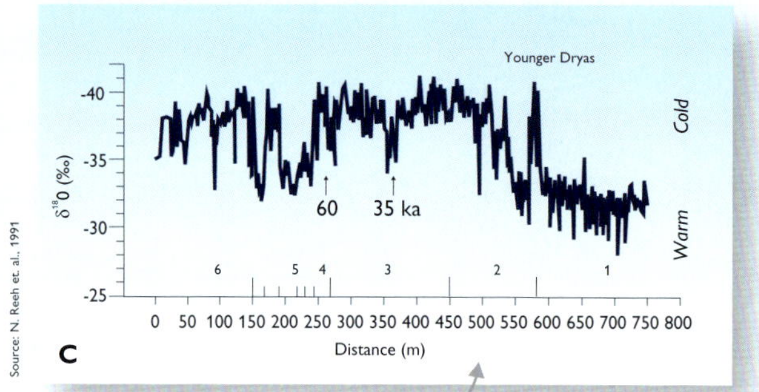

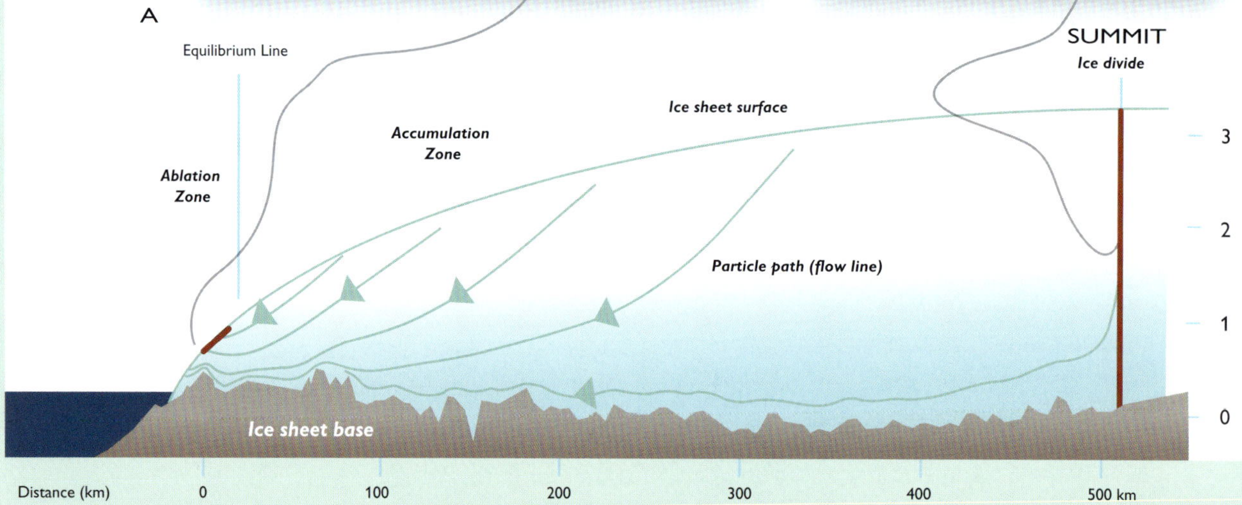

The earliest evidence of a new ice-free period dates from the Kap København Formation, a sedimentary sequence in North Greenland laid down about 2.4 million years ago. At that time, Greenland was without the Inland Ice, which was probably formed during the ice ages that started about one million years ago.

Around Kangia signs of glaciation date mainly from the last ice age, which started 115,000 years ago and ended 11,700 years ago. There are also many traces of the subsequent deglaciation, when the ice margin retreated from the present coastal areas. The retreat ended about 5000 years ago, and since then the ice margin has shown a slight advance. The cool period culminated in the Little Ice Age 500-100 years ago, when the ice margin reached its most recent maximum. The expansion occurred in a series of waves, each pulse resulting in an increase in the area covered by ice.

Glaciation history in the Ilulissat area

During the last ice age the Inland Ice probably covered the continental shelf and the whole of the Ilulissat area. Loss from the margin of the ice sheet occurred by way of major ice streams that produced calf ice. One of the largest ice streams occupied a trough more than 800 metres deep, between Qeqertarsuaq and Aasiaat in the south-west part of Disko Bugt.

The margin of the ice sheet started to break up and retreat in the period from 17,000 to 12,000 years ago. For a 1000 years or more, starting *c.* 10,000 years ago, the margin of the ice sheet was quite stable, and this standstill may in part have been determined by climate. However, evidence from the moraines at the mouth of Kangia indicate that the retreat of the ice margin slowed because a large part of the ice margin was then resting on land and did not calve into Disko Bugt. The subsequent recession reached a maximum 5000-4000 years ago, when the front of Sermeq Kujalleq was 15-20 kilometres east of its present position.

The following colder period, perhaps associated with increased precipitation, led to a re-advance of the ice margin, which culminated in the maximum position of the ice in the 19th century. More recently many of the glaciers have again started to retreat. The ice front of Sermeq Kujalleq retreated 26 km between 1851 and 1950. Strong retreat began again after 2002.

The Inland Ice from top to margin

In the central part of the Inland Ice the accumulated snow is gradually buried, and as it sinks deeper and deeper into the ice sheet, the increasing pressure

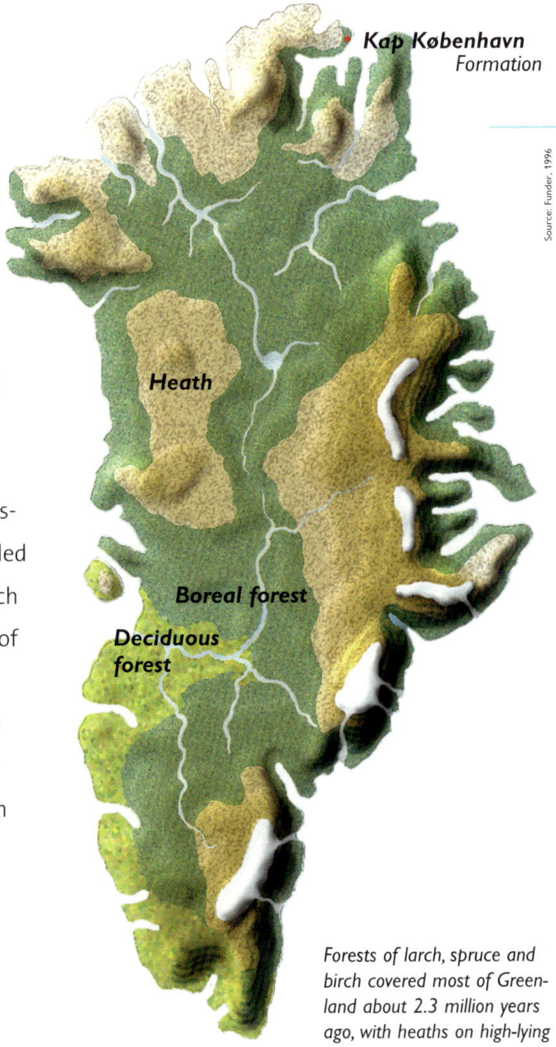

Forests of larch, spruce and birch covered most of Greenland about 2.3 million years ago, with heaths on high-lying areas.

ICEBERGS ARE ALWAYS VISIBLE FROM THE COAST

transforms the snow into glacier ice. This transformation takes place in the upper hundred metres of the ice sheet. Concurrent with the vertical movement downwards through the ice sheet, the ice also very slowly moves down slope towards the sides of the ice sheet.

In general terms this movement starts from the ice divide in the centre of the Inland Ice, and ends at its outermost margins. From the ice divide out towards the ice margin a series of zones are recognised, depending on the altitude of the surface of the ice sheet above sea level.

The highest zone of the Inland Ice is dry, and snow accumulates without any melting. At a lower level there is a zone with some melting and percolation of melt water. In the next zone, the snow can become waterlogged in the course of the summer, but the percolating water refreezes in the deeper-lying snow layers. Closer to the ice margin the refrozen snow creates a new zone at the snow line, which in the Ilulissat area is 1300 metres above sea level; here, the refrozen melt water is added to the glacier ice. This zone ends 1200 metres above sea level at the equilibrium line, where the addition of snow and frozen melt water equals the net loss through melting. The lowest level, extending out to the boundary with the land, is the ablation zone, where melting exceeds the net gain.

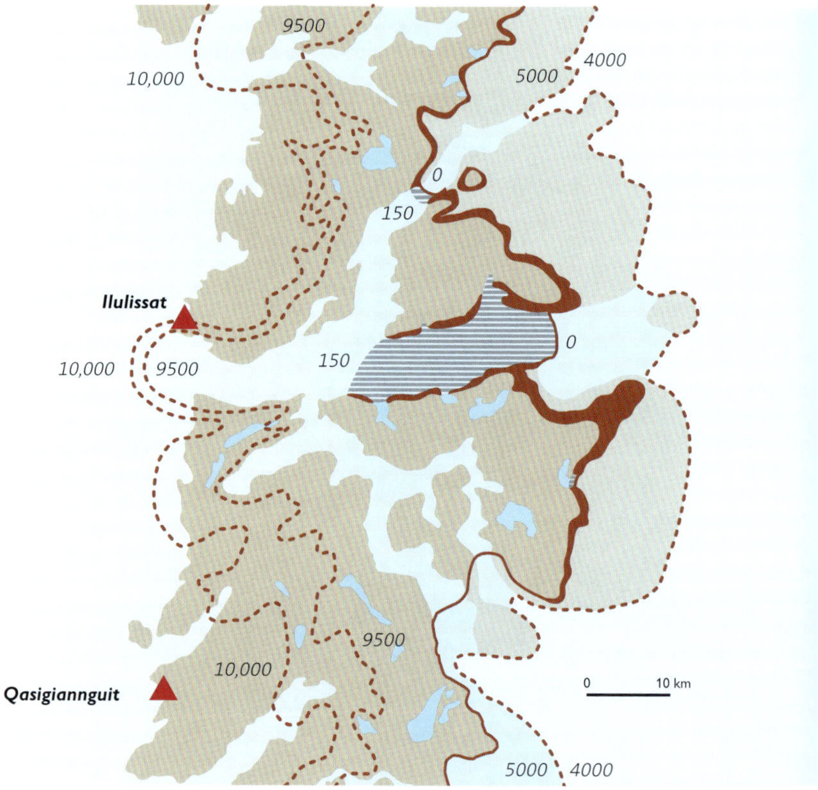

Reconstruction of ice margin positions in the Ilulissat area 10,000, 9500, 5000-4000 and 150 years ago. The change in the ice margin between the Little Ice Age maximum and the current extent of the ice is shown in brown for land areas and shaded for the floating part of Sermeq Kujalleq.

At the summit of the Inland Ice the movement of the ice is slow and is mainly downwards. The degree of sideways movement increases the further one moves towards the ice margin. Outside the limits of the ice streams, the average horizontal flow of the ice reaches velocities of 50-200 metres per year near the equilibrium line, after which the movement falls to near zero at the ice margin.

Portrait of Sermeq Kujalleq

Sermeq Kujalleq can be characterised as a river of crevassed ice. The catchment area that feeds Sermeq Kujalleq is estimated to be 6.5% of the area of the Inland Ice, or 110,000 square kilometres.

The actual ice stream appears as a narrow, 3-6 km wide, well-defined channel that can be followed from the present glacier front about 80 km into the ice sheet. At an altitude of 1350-1400 m above sea level the ice stream splits into three branches, which gradually disappear into the ordinary surface of the ice sheet. Following the course of the ice stream downwards, the surface is relatively smooth until about 50 km from the glacier front, when it gradually becomes more and more crevassed, and lakes and water-filled crevasses disappear. The ice stream has fairly sharp borders against the less-fractured

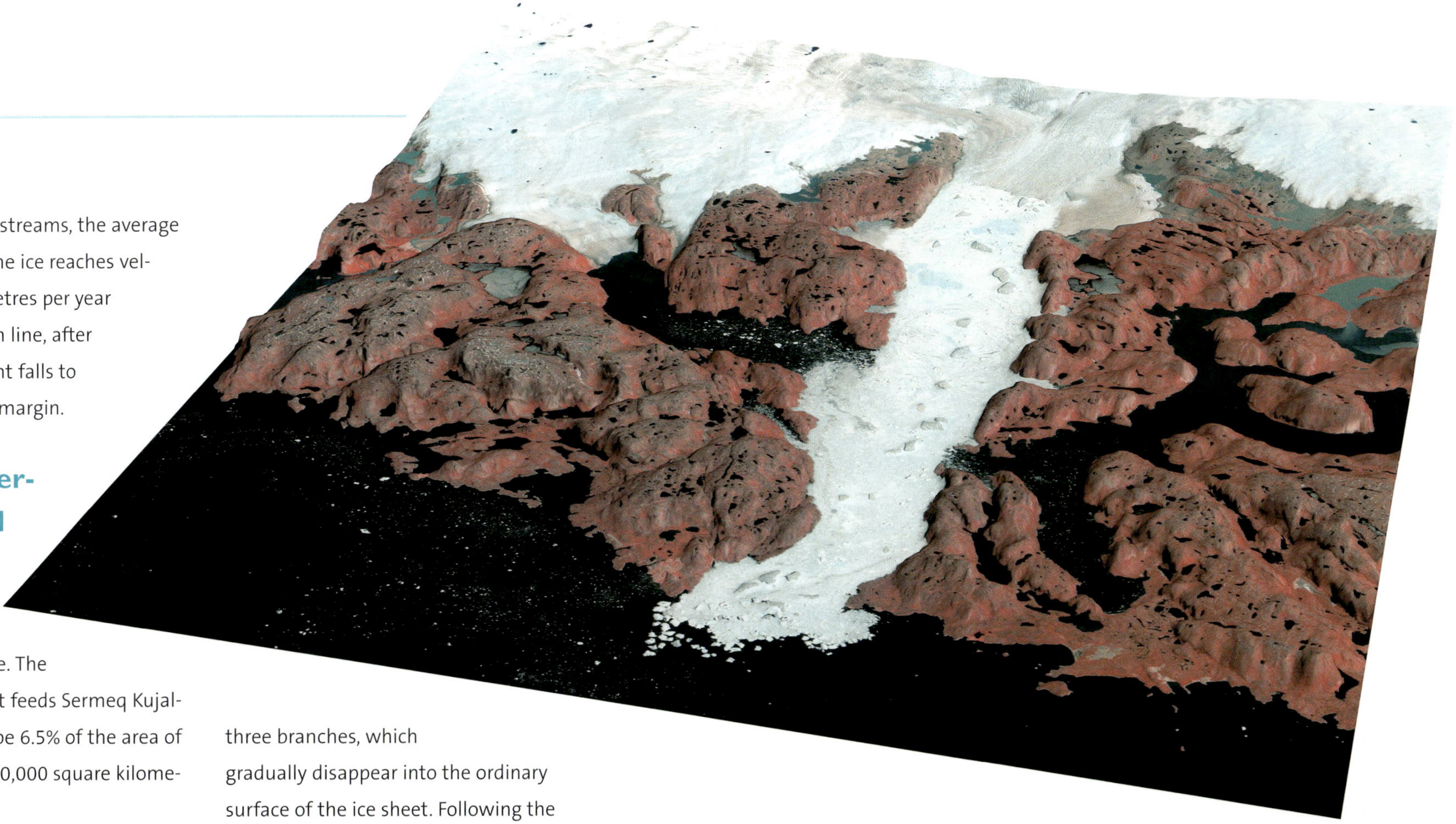

Satellite image of Kangia and Sermeq Kujalleq from 7th July 2001, before the 2002 break-up of the floating front. Vertical scale slightly exaggerated.

MELTWATER STREAM

Undulating part of the ice margin with low ice velocity and few crevasses. Summer surface melting in the ablation zone produces fast-flowing meltwater rivers, which disappear into moulins or holes to join drainage systems beneath the ice surface.

MELTWATER LAKE

Meltwater lakes are found on the ice sheet at altitudes of 1150-1450 m, especially in areas where there are few crevasses. At lower altitudes the ice sheet is thinner, and crevasses are more common. Lakes without drainage systems may develop in hollows, for example at the boundary between the ice stream and the surrounding ice sheet.

CREVASSES

Crevasses can be dangerous if covered by snow.

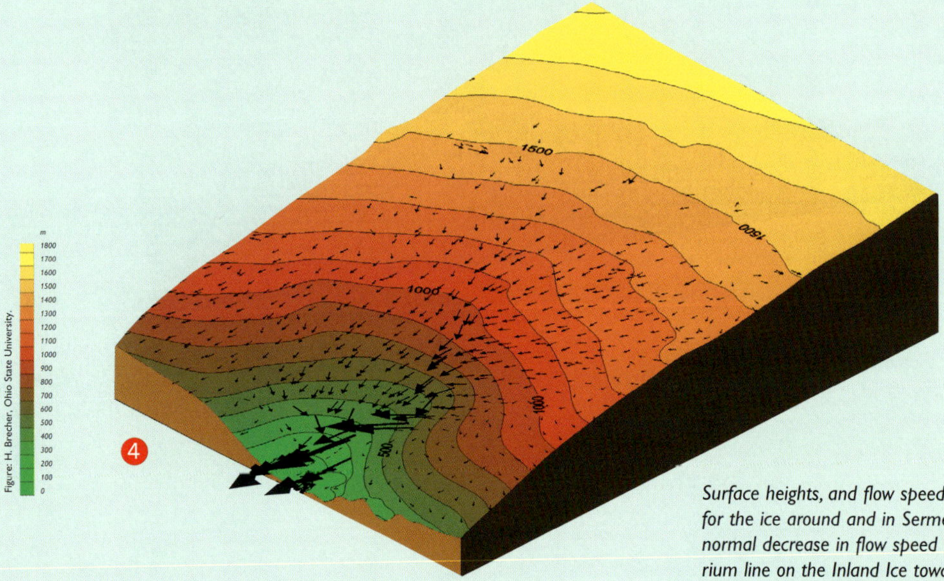

Surface heights, and flow speeds and directions, for the ice around and in Sermeq Kujalleq. The normal decrease in flow speed from the equilibrium line on the Inland Ice towards the ice margin is not seen in the ice stream. In contrast, a dramatic increase in the flow speed occurs, and the surface of the ice changes from smooth and undulating to the chaotic, broken and pinnacled surface of the ice stream.

Sermeq Kujalleq and Kangia. The blue arrows show the movements of the ice stream and the glacier. Vertical scale slightly exaggerated.

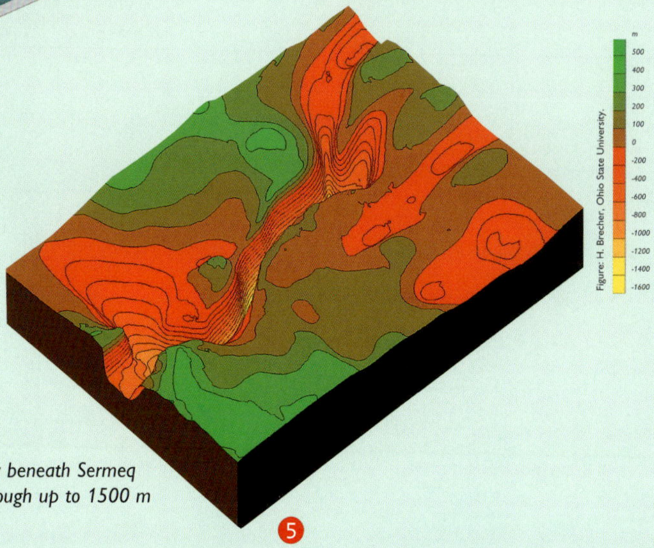

Bedrock topography beneath Sermeq Kujalleq, showing trough up to 1500 m deep.

STRANDED ICEBERGS

Drama in the icefjord.

CALVING FRONT

Sermeq Kujalleq's pinnacled surface.

THE DRAMATIC GLACIER FRONT

surrounding ice. The marginal crevasse zone of the main stream extends 5 km or more on both sides of the ice stream. About 45 km inland the crevasses and the pattern of flow lines show a funnel-shaped convergence of movement towards the main stream of ice.

Prior to the sudden retreat of the glacier front in 2002, a smaller ice stream flowed south into Sermeq Kujalleq about 10 km behind the glacier front. This northerly branch was separated from the main ice stream by an ice rumple that probably rested on the part of a barrier under the ice. This barrier may have contributed to the prolonged and stable location of the glacier front since the 1950s.

Today we have a reasonably good idea of the topography beneath the ice. The ice stream occupies a very deep trough, and is 1900 m thick at the grounding line, the point where the ice stream loses contact with its substratum and begins to float. As the ice surface is here 400 m above sea level, about 1500 m is below sea level. About 40 km behind the grounding line the ice stream is 2500 m thick, and since the surface of the ice is 1000 m above sea level, the bottom of the trough is 1500 m below sea level. About 50 km east of the grounding line the trough flattens out, and 120 km east of the grounding line no marked trough is discernible on radar profiles.

Source: LANDSAT

Satellite image of the calving front of Sermeq Kujalleq before the 2002 break-up. The 7.5 km long front is 40-90 m high, and reflects a glacier thickness of between 500 and 800 m. Icebergs are seen near the left side of the picture. The ice stream is characterised by linear features. A tributary from the north can be seen joining the main ice stream.

Photo: Henrik Højmark Thomsen

Photograph from a helicopter showing the tributary (at left) joining the main ice stream, with an ice rumple in the background.

Before the 2002 glacier collapse, the outermost 12 km or so of Sermeq Kujalleq were afloat. The floating segment of the glacier partially fill the inner part of Kangia and moves up and down with the tide, which has a maximum height difference of 3 metres; the tidal effect falls off towards the grounding line. As the tide rises and falls the actual grounding line changes slightly in position, such that there is in effect a grounding zone. Movements of the ice, felt as tremors, vary in strength with the tidal cycle, and can be detected up to 8 km inland from the grounding zone.

Calving

Estimates of the annual production of calf ice from Sermeq Kujalleq rely primarily on knowledge of the flow velocity of the ice and the thickness and width of the calving front. Calculations of the calf ice production vary greatly, but a reasonable estimate is about 50 cubic kilometres of ice per year.

Major calving events, in which large segments of the glacier front detach, take place only a few times in the course of the summer. On the other hand, there is almost continuous calving of smaller pieces of ice from the ice front, and this can be observed all year round.

The process of calving is not yet fully understood. The most important trigger is probably the expansion of the crevasses in the bottom of the glacier, which develop as the glacier detaches from its substratum and begins to float in the fjord. At the same time the glacier is deformed by the effects of the tide on the floating segment, and the tide also induces decoupling of the glacier from the cliffs on both sides of the fjord. The pressure of water filling the glacier crevasses further strengthens the loosening process.

BOX 4: KANELING

Fishing vessels in Ilulissat harbour.

Calving icebergs once caused waves in Ilulissat harbour

In earlier times, a strange phenomenon - 'kaneling' - could create problems in Ilulissat harbour.

This is a description from 1921: *"The water in the harbour suddenly becomes quite rough, and can even become foaming white, and grass and seaweed from the bottom come up to the surface. However, there will rarely be any danger for ships or boats there as long as their moorings are solid. It all lasts only a few minutes before calming down again. In recent years, 'kaneling' seems to be on the decline and must now be said to be quite without importance".*

The phenomenon was investigated when a new pier was constructed in Ilulissat harbour. 'Kaneling' is related to pressure waves, formed during calving at the ice front, and which become stronger because of resonance in the harbour basin. That the problem declined at the beginning of the 20th century was probably due to the considerable retreat and thinning of Sermeq Kujalleq.

When large icebergs overturn in Disko Bugt, major waves are formed that can be dangerous for boats in the bay and for people on the coast. At Sermermiut, tourists have been inundated by waves from overturning icebergs. This type of wave is not related to 'kaneling'.

Another phenomenon connected with the iceberg bank (Isfjeldsbanken) was also described in 1921: the formation of an ice-dam across the iceberg bank. The formation of such a dam can cause the water-level in the icefjord to rise above the water level in Disko Bugt. According to the description, the ice-dam breaks when it can no longer withstand the pressure from the dammed waters in the icefjord, and 'the icefjord shoots out'.

Photo: Anker Weidick

SERMEQ KUJALLEQ SEEN FROM THE SOUTH SIDE OF THE ICEFJORD

In the period from March to October 1982, calving and iceberg production was closely followed using data from satellite remote sensing. Three major calving events occurred. The largest produced 14 tabular icebergs, with sizes up to 2 km by 1 km. This event resulted in the estimated discharge of 8.4 cubic kilometres of ice and a retreat of the glacier front by about 2 km, and was associated with the break-up of the sea ice in the in-

Calving at the glacier front.

Photo: Dieter Zillmann/Greenland Tours Elke Me

ner part of Kangia. Another pronounced calving, with the detachment of large tabular icebergs, was observed in July 1985 when the outer 2 km of the floating part of Sermeq Kujalleq disintegrated in the course of just 45 minutes.

After calving in a normal summer, the glacier front gradually advances during the winter, since the front is supported and stabilised by the sea ice in the fjord. The seasonal movements of the glacier front were studied in the period 1962 to 1996, and the typical difference between the frontal positions in the summer and the winter was 2-3 km.

Dynamics of Sermeq Kujalleq

Glaciers normally move down the steepest surface slope, and it is thus the topography under the ice that determines the course of valley glaciers in mountainous regions all over the world. However, the

BOX 5: BUBBLES AND GLACIER ICE

Ice can be formed in two ways:

- by freezing water or
- by compression of snow.

Greenland's Inland Ice is composed almost exclusively of glacier ice, formed by the compression of snow.

A pronounced characteristic of glacier ice is the granular structure, with ice grains of different sizes interlocked into each other like pieces of a jigsaw puzzle. The crystals are formed from the original snow crystals, that in the course of time have grown under pressure at the expense of neighbouring crystals. During the process the air in the snow is trapped as bubbles in the ice. A study of the icebergs released from Greenland glaciers shows that the majority of air bubbles have a pressure of 4-5 atmospheres, but pressures up to 20 atmospheres have been measured.

The white colour of the ice is due to the bubbles. Blue or green bands of pure ice that are frequent in icebergs were formed by freezing of bubble-free melt water in the former crevasses in the glacier.

The generally white colour of the ice is due to the presence of air bubbles. Blue or green bands, often seen in icebergs, are formed by frozen meltwater that filled former glacier crevasses.

BRASH ICE BETWEEN FLOATING BERGS

ice sheets in the polar regions are not forced to follow the underlying topography, but instead overrun entire landscapes and create their own pattern of movement. The flow of the ice is also directed downwards here, but it is controlled in particular by the surface topography of the ice sheet, and the flow pattern does not necessarily follow the underlying topography.

In most of the marginal areas of ice sheets the ice moves slowly, and there is little variation in lateral velocity. In other marginal areas the flow can be faster and concentrated in stream-like channels. Tongues of ice that extend down into valleys and fjords are called outlet glaciers, no matter whether they end on land or in water. When a glacier runs into a fjord or a lake, the flow velocity normally increases because of the reduced friction at the bottom. The large ice sheets in Greenland and in Antarctica have outlet glaciers that can locally move at velocities of several kilometres per year; this happens when the glaciers discharge through deep troughs where the basal friction is reduced.

Sermeq Kujalleq is a classic example of an ice stream and its high velocity is due to ice from a large drainage area becoming concentrated into a narrow stream. The crucial question is the nature of the climatic conditions that create a permanent fast-flowing ice stream. The large production of calf ice and the high velocity indicate that Sermeq Kujalleq reacts quickly to climatic changes, but the reasons for the permanently high rate of flow are still under debate.

Traditionally, the term used by glaciologists to explain the high discharge of ice is known as the 'Jakobshavn effect'. This term derives from the now obsolete Danish name 'Jakobshavn Isbræ' formerly used for Sermeq Kujalleq glacier. The 'Jakobshavn effect' appears to be a combination of crevasse formation due to increased melting, and increasing acceleration of the disintegrated current of ice. Surface melting is greatly enhanced by extensive surface crevasses. When large quantities of melt water freeze internally in the glacier, considerable latent heat is released and this makes the ice softer and more fluid. Meltwater that reaches the bottom of the glacier acts as a lubricant that helps the ice masses to flow more quickly over the underlying rock.

It has been suggested that the movement of Sermeq Kujalleq started to accelerate about 1850, when rising temperatures after the end of the Little Ice Age led to increased surface melting of the lower parts of the ice sheet. Meltwater that ran down into crevasses and moulins, may have warmed the ice and lubricated the bedrock surface, and thus initiated Sermeq Kujalleq's fast rate of flow. The high velocity in its turn contributed to the disintegration of the ice into a chaotic mass, so that the surface area exposed to melting became three times as large. The consequences were increased melting, more meltwater to heat up the ice and lubricate the underlying bedrock, and increased transport of ice from the inner part of the Inland Ice towards the fjord. However, there are no records of the glacier's movements before 1875, and the earliest historical descriptions of the iceberg bank indicate there was a high production of icebergs long before 1850. Whether the acceleration of the glacier began about 1850 is thus an open question.

Classic glaciological theory suggests the 'Jakobshavn effect' sustains Sermeq Kujalleq's permanent high speed. However, other proposals question the necessity of this effect to explain Sermeq Kujalleq's continued fast flow.

Dramatic break-up in 2002

In the autumn of 2002, a number of detached parts of the glacier front were observed at Isfjeldsbanken; their characteristic feature was a dark, pinnacled surface. The quantity of such fragments was particularly striking in the summer of 2003, and flights over Kangia and Sermeq Kujalleq showed that the fjord was closely packed with broken up segments of the glacier front. The boundary between these fragments and the glacier front was at times difficult to determine precisely, unlike the situation during the 'stable' period of the last half-century, when the frontal position of the glacier in Kangia was usually clearly defined.

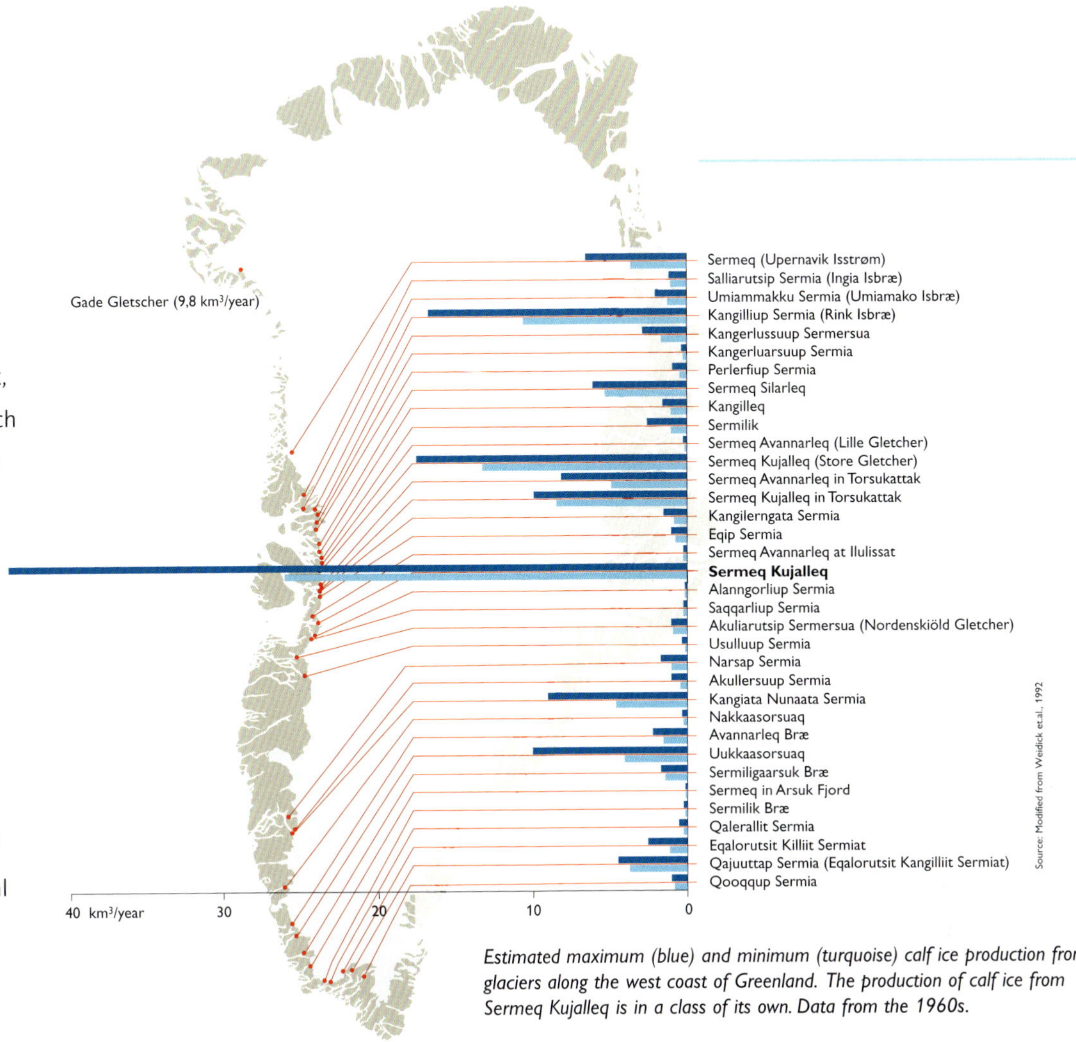

Estimated maximum (blue) and minimum (turquoise) calf ice production from glaciers along the west coast of Greenland. The production of calf ice from Sermeq Kujalleq is in a class of its own. Data from the 1960s.

Studies of satellite images have enabled the position of the glacier front to be determined for the period May 2001 to May 2003, supplemented by data from an aircraft that flew over the area in August 2003. The images show that the last winter advance of the glacier tongue occurred about March 2002, and that the subsequent disintegration of the floating glacier tongue has resulted in the position of the glacier front retreating 12 km eastwards.

WASHBOARD OF ICE

The large ice bay, Tissarissoq, on the southern side of the former glacier front, is now isolated from the main glacier and was showing signs of extensive melting in the summer of 2003. Numerous finds of marine bivalve shells in the moraines around the bay show that Tissarissoq was formerly a marine bay, and it may be that the marine bay may redevelop once again.

The icefjord

Kangia has only been investigated to a limited extent. However, there is no doubt that the inflow of melt water and the ice from Sermeq Kujalleq have a crucial influence on the salinity, currents and temperatures in the fjord. The large freshwater input from the ice sheet and the melting icebergs also has a major influence on the waters beyond the icefjord.

The size of the icebergs indicates that the deepest part of the fjord is between the glacier front in the east and the iceberg bank in the west at the mouth of the fjord. Some measurements made from the sea ice in the fjord indicate depths of over 1000 m in the central parts, but systematic depth profiles along the fjord have not yet been made. Kangia is normally full of ice the whole year round, and the seasonal variations in the ice cover have been studied in detail using satellite images. From June to September the fjord is normally filled with brash ice and drifting icebergs, while sea ice covers the water between the icebergs the remainder of the year. In recent years, however, the outer parts of the fjord have often been ice-free for long periods in the autumn and early winter. The steady advance of the glacier front, however, means that cracks are continually forming and re-freezing in the sea ice during the autumn, winter and spring.

The iceberg bank

Isfjeldsbanken, the iceberg bank, lies at the mouth of the icefjord. Icebergs that are calved from the front of Sermeq Kuj-

Satellite image of Sermeq Kujalleq and Kangia from 7th July 2001, before the 2002 break-up of the floating front. Data from the Landsat-7 satellite.

The colours are artificial. Areas with vegetation are brown or red. Barren rock surfaces are grey.

The glacier front is situated at the position where it was between the 1950s to the break-up in 2002.

Satellite image from 28th May, 2003, showing the front of Sermeq Kujalleq after the 2002 break-up. After the break-up of the floating front, the new glacier front is situated 12 km further east. Data from the ASTER satellite.

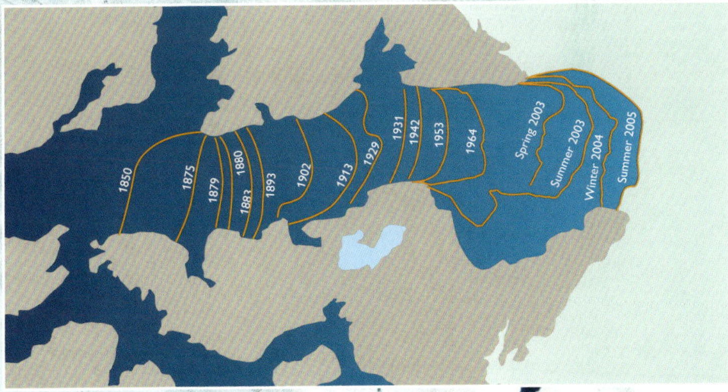

The development in the front of Sermeq Kujalleq between 1850 and 2005.

FOG MAY SUDDENLY ENTER THE ICEFJORD

alleq drift slowly towards the mouth of the icefjord, a process that takes between 12 and 15 months. The depth of water over Isfjeldsbanken is only about 200-225 m, and here most of the larger icebergs run aground.

The iceberg bank is a curved, low-water threshold at the mouth of Kangia, and stretches from Sermermiut in the north to Avannarliit in the south. There are moraines on each side of the fjord, which are composed of pebbles and boulders and formed at the margin of the Inland Ice 10,000-8000 years ago. It is assumed that the line of moraines formerly crossed the mouth of Ilulissat Icefjord and formed the top of the iceberg bank. The morainic sediments have probably been largely eroded by the icebergs subsequently, and it is probable that a sill of bedrock occurs beneath the remnants of the moraine, as is the case in other threshold fjords in Greenland.

A few soundings at Isfjeldsbanken, given on nautical charts, show depths of 200-225 metres. This shallow threshold separates the deep trough of the icefjord from the 200-400 m deep sea bottom in the inner part of Disko Bugt. Further towards the south-west a deep trench known as Egedesminde Dyb is found, where water depths exceed 800 metres. This trench was formed by the activity of another ice stream during the ice ages. Large icebergs go aground on the iceberg bank throughout the year. They remain there until they have either melted so much that they can float over the threshold, or the pressure from the ice in the icefjord becomes so great that the icebergs are forced across the threshold and into the deeper water of Disko Bugt.

Large icebergs with a draught of more than 200 metres run aground on Isfjeldsbanken.

BOX 6: ICEBERGS

Photo: Dieter Zillmann/Greenland Tours Elke Meissner

A boat trip between pinnacled icebergs can be an awe-inspiring experience.

Source: Wadhams, 2002

Distribution of icebergs in the North Atlantic, and reports (with year) of exceptional sightings. The sinking of Titanic in 1912 occurred south-east of Newfoundland.

The large icebergs that ground on Isfjeldsbanken at the mouth of the icefjord act as a barrier that causes smaller icebergs to pile up in the fjord. The icebergs are very different in size and form. They vary from small pieces of ice to giants that tower more than 100 m above sea level. Some icebergs are rounded, while others have extremely irregular shapes, often jagged and with pointed summits.

It is often said that only a tenth of an iceberg is above sea level. However, glacier ice contains many air bubbles and the part visible above the water is really about one-ninth of the total volume. The content of air bubbles or moraine material can also affect the depth of an individual iceberg. Icebergs with a heavy load of gravel and boulders can be so heavy that only a small part projects over the water. In addition, water salinity plays a role in deciding the depth at which an iceberg floats. The saltier the water, the higher an iceberg will project above the water surface.

Melting and erosion starts as soon as an iceberg is released from the front of the glacier, but usually progresses slowly in arctic waters. Warm air and the effect of the sun's rays cause the part of the iceberg that is above water to melt. Beneath the water surface, the iceberg is melted by warm salty water and is eroded by the dashing of waves. The processes occur at different speeds at different places on the iceberg, which can therefore become unstable and overturn.

A danger at sea

When icebergs escape across the threshold (Isfjeldsbanken) at the mouth of the icefjord, they drift further out through Disko Bugt to the open sea. Some icebergs drift south around the island of Disko, while the majority drift north around the island and there enter Davis Strait between Greenland and Canada. Here the icebergs meet the West Greenland Current, which carries them northwards along the Greenland coast. Further north, the icebergs are carried by ocean currents westwards towards Canada, where they are caught by the Baffin and Labrador Currents and drift southwards.

Many of the large icebergs drift as far south as the latitude 40°N before they melt. Icebergs from Kangia, and from other icefjords and major glaciers in west and North-west Greenland, have always been a danger to shipping on the North Atlantic routes, and it was probably one of these icebergs that sealed the fate of the Titanic in 1912. After the catastrophe an international conference decided that an ice patrol should be established to observe and track icebergs. The US Coast Guard started monitoring them in 1914, and the task was expanded in 1928 to include investigations of where the icebergs originated. A total of 21 outlet glaciers in West Greenland were identified as possible sources of icebergs, but only a few of them, and Sermeq Kujalleq in particular, were considered as being important.

The interest in monitoring icebergs fell in the 1950s and 1960s, probably due to a pronounced fall in the number of icebergs encountered compared with the preceding decades. This development changed dramatically in the 1970s, and 1972 and 1974 were the most iceberg-rich years ever registered. At the same time the oil crisis had generated interest in exploiting oil along Canada's coast and offshore West Greenland, which gave iceberg studies a completely new perspective.

STRANDED GIANTS ON ISFJELDSBANKEN

The land below and around the ice

SPELLBOUND VIEW FROM THE COAST

The geology of Greenland reflects a history that stretches back over nearly four billion years. By far the greatest part of the sub-continent is made up of crystalline rocks of the Precambrian shield. Fjords and valleys that cross the present day shield were formed by river erosion and later deepened by glacial erosion during the many advances and retreats of the Inland Ice. The glaciers also erode the landscape today, as can be seen at the ice margin around Ilulissat.

The Precambrian shield of West Greenland consists dominantly of crystalline gneiss basement, formed during several periods of mountain building, and welded together. The oldest rocks about 3.8 billion years old occur in the Isua region to the south, whereas the area around Ilulissat is formed by younger rocks about 1.8-1.6 billion years old.

The Precambrian shield is dominated by pale grey, folded, gneisses and granites, with bands and enclaves of mica schist and dark basic rocks. The bedrock is widely exposed along the whole of the ice-free part of western and southern Greenland, but the Inland Ice conceals the bedrock of 81% of Greenland.

Some of the oldest rocks in the world are found at Isua, north-east of Nuuk, the capital town of Greenland. Here the mountains contain 3.8 billion-year-old rocks that have been intensively investigated by geologists, and which contain the earliest signs of life on Earth in the form of carbon thought to be produced by micro-organisms. Evidence has also been found for plate tectonic processes even at this early time. The ancient rocks around Isua are bounded by younger crust formed 2.5 billion years ago, with to the north and the south still younger rocks formed 1.8 to 1.6 billion years ago. The bedrock between Kangerlussuaq and the inner part of the Disko Bugt region, comprises both 2.5 billion years and 1.8-1.6 billion years old rocks that were formed deep down in the Earth's crust.

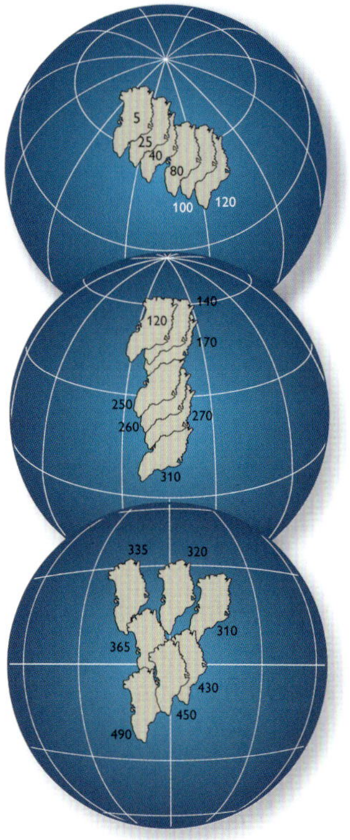

Modified after Gunver Krarup Pedersen, 1999

The geographical position of Greenland has changed systematically over the past 500 million years as a result of plate tectonic movements, from the tropics to its current position in the Arctic. The figures give time in million years before the present.

Together with North America, Greenland formed a large North American continent, 1.6 billion years ago. At that time the folding in the Precambrian mountain ranges in Greenland was more or less complete, and the shield had risen and was partly eroded. Later developments largely involved the marginal parts where rocks formed later are preserved in North Greenland, East Greenland and parts of central West Greenland. Two large fold belts were formed along the coasts of North and North-East Greenland 430-350 million years ago.

Greenland's subsequent geological development was dominated by the formation of sedimentary basins along the margins of the Precambrian shield, where 5-10 kilometre thick successions of sediments were deposited. These deposits are preserved today in coastal areas, with the thickest successions offshore on Greenland's continental shelf. The sedimentary rocks on Disko, and the peninsula of Nuussuaq north of Disko Bugt, were formed 120-60 million years ago. The rock types include deposits from rivers and lakes and marine sediments deposited off the then coast.

Tectonic plate movements 60-55 million years ago led to the opening of the Atlantic, when Greenland was separated from Norway and northern Europe. Meanwhile the Labrador Sea and Davis Strait resulted from movements along a spreading axis between Canada and West Greenland. Sea-floor spreading was accompanied by profuse volcanic activity along the rupture zones, and the resulting km-thick volcanic successions are preserved today on Disko, Nuussuaq and Svartenhuk Halvø in West Greenland, and in the central part of East Greenland. Similar lava-successions are widespread on the sea floor offshore West and East Greenland.

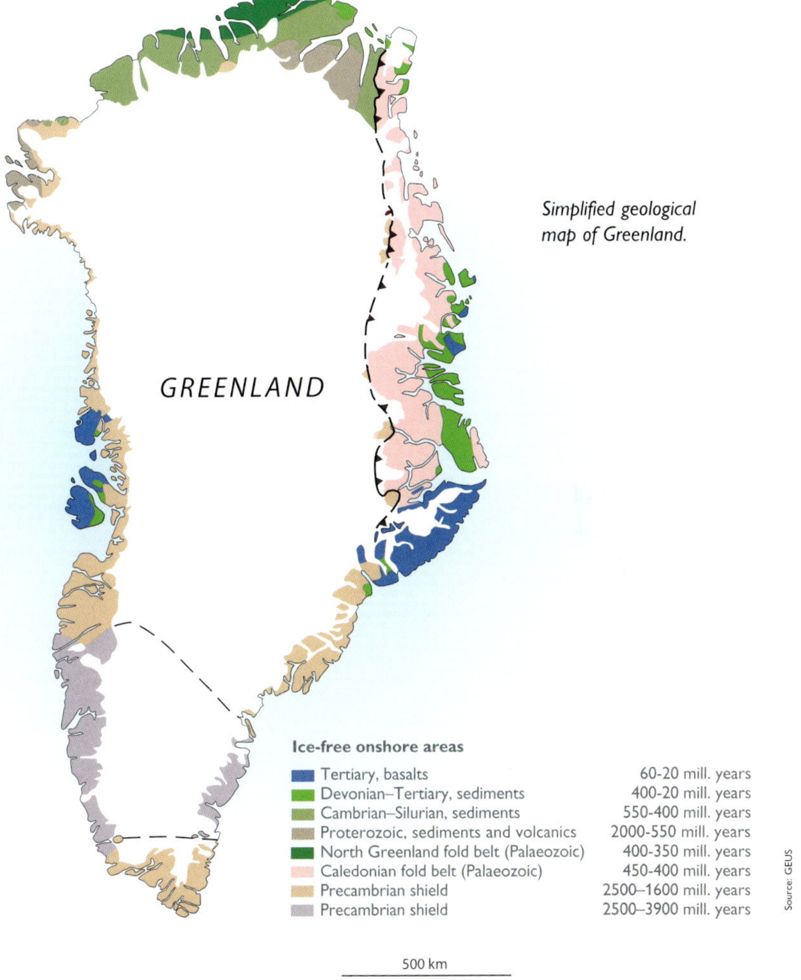

Simplified geological map of Greenland.

Drainage channels, troughs and fjords

Kangia is eroded through an area of Greenland's Precambrian shield formed 1.8 billion years ago. The mineral compositions of the rocks indicate they recrystallised at depths of 20-30 km and temperatures of 400-700°C. Uplift and

ICE PYRAMID ON ITS WAY TO ISFJELDSBANKEN

erosion has brought these once deeper-lying crystalline rocks up to the present day land surface.

Before the Inland Ice covered Greenland, rivers played an important role in the early formation of the valley that became the site of Kangia and the trough under Sermeq Kujalleq. The rivers eroded valleys that extended right out to the edge of the continental shelf. The watershed lay in the eastern part of Greenland, such that a substantial proportion of the precipitation that fell over Greenland drained westwards towards Disko Bugt. River erosion must have led to the formation of major valleys.

As the Inland Ice began to cover Greenland, a large part of the precipitation continued to drain towards Disko Bugt, but now in the form of glacier ice. The major ice streams followed the major valleys that had been formed by the rivers. Large glaciers such as Sermeq Kujalleq can erode broad and deep troughs, where the bottom lies far under sea level. The ice itself is not a very effective erosion agent, but rocks and boulders at the base of the glacier are pushed over the bedrock under the considerable weight of the ice and cause the glacier to act like a gigantic file. It is in this way that the large glaciers around Disko Bugt have formed Egedesminde Dyb, Kangia and the trough under Sermeq Kujalleq.

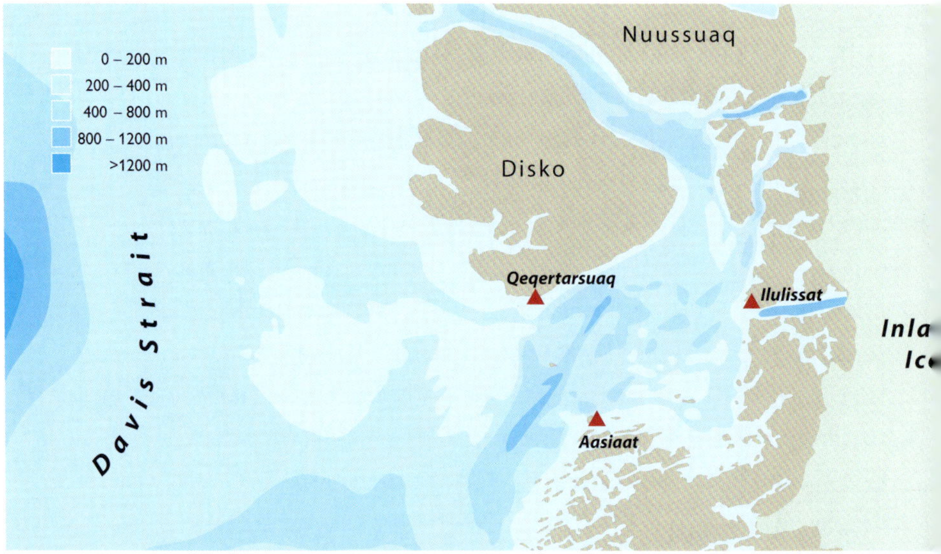

Water depths in Disko Bugt and Davis Strait. The Egedesminde Dyb trench between Qeqertarsuaq and Aasiaat marks the position of a former ice stream.

Glaciers are responsible for creating many of the spectacular fjords around the world. The Ilulissat area is particularly important as one of the few localities on Earth where the power of active glaciers in eroding fjords and the potential of meltwater forming drainage channels can be seen at the present day.

Moraine ridges, kame terraces and marginal deltas

A large part of the rock material eroded from the trough beneath Sermeq Kujalleq and Kangia has been deposited on the sea floor over thousands of years. There are also moraine ridges on land, and a well-developed system of moraine ridges can be traced over the land from one fjord to the next, stretching almost without a break from Qasigiannguit and Ilimanaq in the south via the iceberg bank to the mouth of Paakitsoq fjord.

The earliest moraines were deposited 10,000 years ago, when the relative sea level was 60-70 metres above the present day sea level. A later moraine system formed 8000 years ago, when the relative sea level was 35-50 metres above that of the present day. The disposition of the moraines and the shallow water depths over Isfjeldsbanken indicate that the calving front of the forerunner of Sermeq Kujalleq was considerably smaller then than it is today, and the production of calf ice must have been similarly smaller.

Moraine ridges are formed during pauses in the retreat of the ice front or during short re-advances. Features such as kame terraces and marginal deltas are also formed at the ice margin. All three landscape forms can be seen in areas around the present ice margin near Ilulissat, areas from which the ice has disappeared over the course of the past 150 years.

Land uplift

When an ice sheet disappears, the land rises. Signs of land uplift can be seen at many places in Europe and North America. The Swedish 'Höga Kusten' is an exceptional example. During the ice ages, Scandinavia and large parts of northern Europe were covered by the Fennoscandian ice sheet. The ice sheet was three kilometres thick and depressed the Earth's crust in central Scandinavia by at least 800 metres.

Greenland is a present-day analogy to the Swedish 'Höga Kusten', in that the three-kilometre thick Inland Ice depresses the crust below the inner part of Greenland to about the same extent as central Scandinavia during the last ice

Landscape (c. 600 m a.s.l.) at Tasersuaq lake, between Sikuijuitsoq and Paakitsoq fjords, viewed towards the North-west. The moraine, with a person standing on the top, is part of a system of ice margin features that extend from Isfjeldsbanken to the mouth of Paakitsoq. In the right background the margin of the Inland Ice is visible at the head of Paakitsup Ilorlia. This was the location of hydro-power investigations for the town of Ilulissat town in the 1980s. In 1886, Robert E. Peary and Christian Majgaard visited the ice margin in this area.

Photo: R. Bøgvad 24th July 1961

ICEBERG WITH A HEAVY LOAD OF MUD AND GRAVEL

age. The area around Kangia is reminiscent of Sweden's 'High Coast' during the ice age, apart from the fact that the present ice-free coasts have undergone uplift following the retreat of the Inland Ice margin from its former maximum extent.

The last ice age culminated 21,000 years ago, and the world's ice sheets have subsequently either disappeared or diminished in size, with the consequent contribution of meltwater causing the level of the world's oceans to rise by 130 metres. Areas of Greenland that were once covered by ice have been raised above sea level due to uplift. In West Greenland, for example, the presence of raised marine and coastal deposits in the present day ice-free land areas are evidence of land uplift.

In the western part of the Ilulissat area the highest coastline dating from the end of the ice age is 70 m above the present sea level. There are raised marine deposits in several places, for example at Sermermiut, where sand deposits contain shells of marine bivalves.

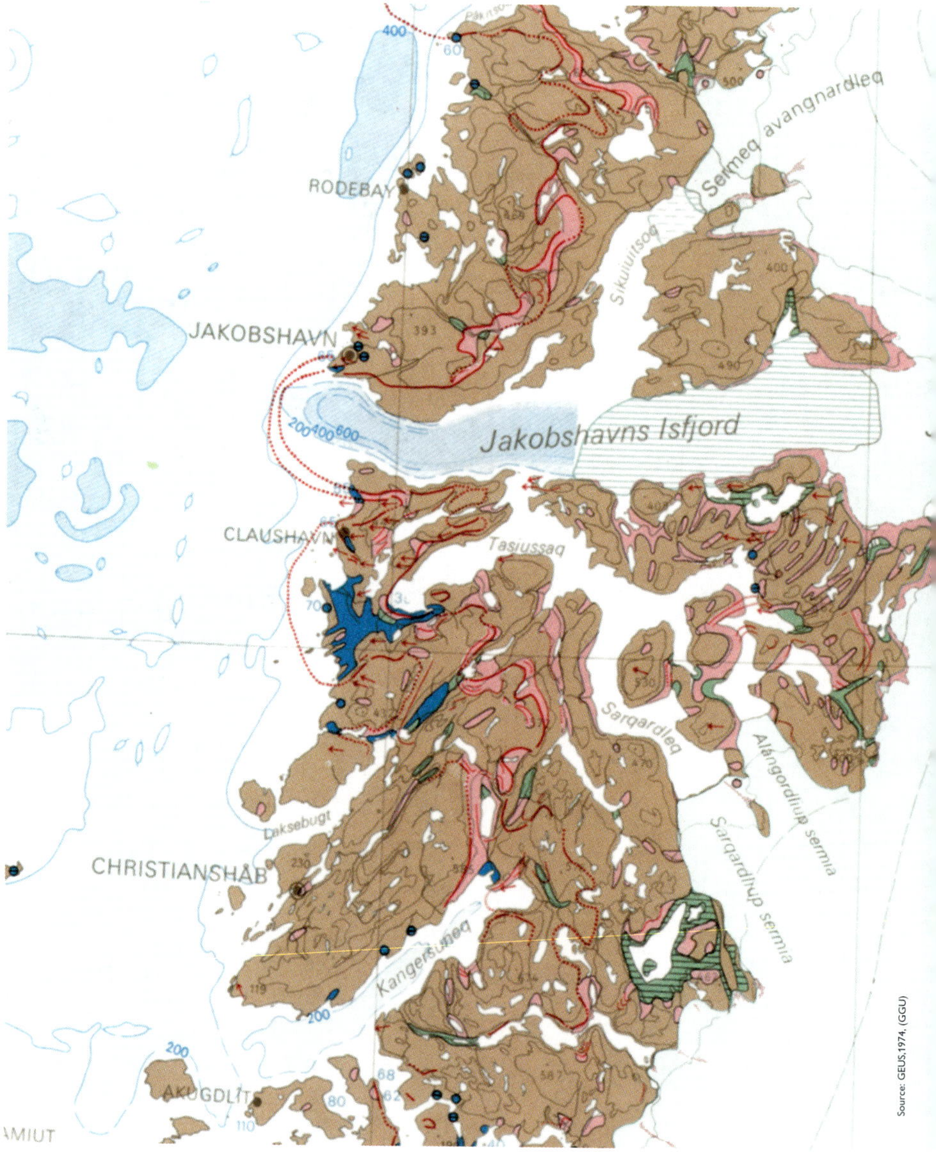

Major moraines (red lines) in the Ilulissat region. The moraines related to ice recession are shown in pink. Marine deposits are shown in blue and lake sediments in green. The former (19th century) extent of Sermeq Kujalleq (Jakobshavn Isbræ) is shown with blue-grey horizontal lines. Marine deposits are shown in blue and freshwater deposits in green. Reproduced from a published 1974 map sheet, that uses the former Danish names for Qasigiannguit (Christianshåb), Ilimanaq (Claushavn), Ilulissat (Jakobshavn), Kangia (Jakobshavn Isfjord) and Sermeq Kujalleq (Jakobshavn Isbræ).

Source: GEUS,1974, (GGU)

Floating icebergs sail towards the open sea.

BOX 7: WHEN THE ICE ADVANCES AND RETREATS

The repeated advances and retreats of the ice sheet margin have created many small-scale landscape features.

ICE-SCOURED BEDROCK WITH PERCHED BOULDERS.

- Many of the valleys in the area around Ilulissat are glacially modified river valleys, like the icefjord itself. The valleys are formed in zones where rocks with many joints and fractures can more easily be eroded by waterand ice.
- Roches moutonées (literally 'sheep rock') are asymmetrical bedrock mounds, with a smooth rounded stoss side and a steep rough lee side. They are formed when the ice slides over the terrain, grinding to form the rounded side, while the rough side is formed when the ice plucks stones and boulders out of the bedrock.
- A characteristic feature of the area around Kangia are the numerous lakes that lie in hollows formed by glacial erosion. Their great concentration and random location reflect the strongly irregular relief.
- Erosion marks such as glacial striae are formed by stones embedded in the base of the ice sheet, and are formed as the ice moves so that the striae are parallel to the movement direction of the ice. These striations were presumably formed during the period prior to the last deglaciation of the area, which took place around 10,000 - 5000 years ago.
- So-called perched boulders are conspicuous in many places. These are large blocks or boulders that rest in an often unstable position on rock surfaces, where the retreating, melting ice left them. They are not usually found in areas close to sea level, where breaking waves can disturb them.
- Newer ice front features include the so called 'trim lines', on rock surfaces that mark the surface of an earlier glacier. Beneath the trim line the rock surface is light and without lichen growth, while the rock above the line (not covered by the earlier glacier) is darker and covered by lichen and other plants.

North side of the front of the glacier, with trim line. The light coloured vegetation-poor zone was formerly covered by an extension of the glacier. Dark coloured rocks in the background (above the trim line) are covered with lichens and other plants.

Climate, darkness and light

Ilulissat lies north of the Arctic Circle. In winter the sun does not rise above the horizon from November to February (polar night), and similarly in summer the sun does not set between early May and August (midnight sun). The sun is always low in the sky in the polar regions, and this is the reason for them being amongst the Earth's coldest regions.

Ilulissat is characterised by a continental climate, although the town lies on the coast. It lies in the low arctic climate zone, which is defined as having an average temperature during the year's warmest month of between 5 and 10°C. The warmest month is July, and the average temperature for Ilulissat in July during the period 1961 to 1979 was 8.3°C. The average daily maximum temperature for July was 11.4°C. Ilulissat has low temperatures in winter and the coldest month is March, with a mean temperature of minus 14.7°C. The lowest average daily minimum temperature that has been registered at Ilulissat is minus 19.3°C.

Normally there is high pressure over the Inland Ice. In combination with the high altitudes of central Greenland, the high pressure over the ice sheet means that low-pressure systems are often forced to move northwards along the western coastal areas. Most of the precipitation in West Greenland is due to the passage of low-pressure systems. At Ilulissat the average annual precipitation is only 266 mm, with August and September as the wettest months.

Calm days are quite frequent, especially in summer, but the weather can also be very stormy. The highest wind speeds are recorded in winter, and the highest average speed of 22.1 m/s was registered in January 1992. Storms occur when low-pressure systems pass Ilulissat, or can be associated with the relatively warm and dry föhn winds, which blow from the Inland Ice westwards towards the coast.

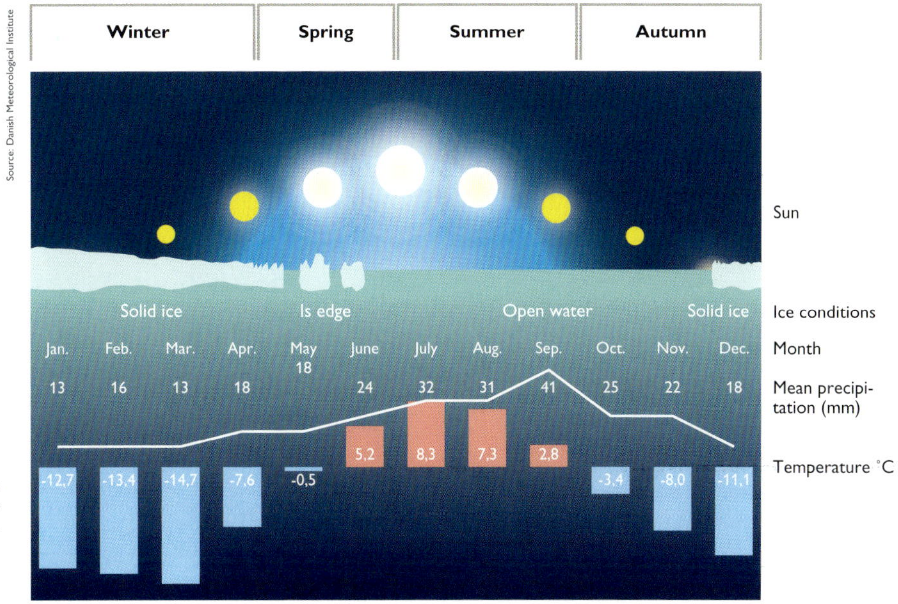

Distribution of sunlight, sea ice conditions, temperature and precipitation during the year in the Disko Bugt area.

During föhn winds the temperature can rise by 10-20°C within a few hours, and the air often becomes unusually dry. In winter, the warm föhn winds can cause snow and ice to thaw over large areas, but föhn winds usually do not last very long, and the normal winter weather soon returns with its biting cold. Refreezing after föhn winds often results in formation of a crust of ice on the snow surface.

IN THE SUMMER, THE SUN SHINES ALL DAY

A "POLAR BEAR" OF ICE

Systematic meteorological observations started in Ilulissat as early as 1873. The continuous series of records since then document pronounced climate changes. Between 1900 and 1925 the mean winter temperature rose by 4.5°C. Summer temperatures also increased during this period, but to a lesser extent. The warming is attributed in particular to an influx of warm Atlantic air masses from the south, but the temperature increase has been intensified through a reduced albedo effect, because of a reduced extent of the sea ice. The white sea ice reflects sunlight back into space, whereas in the absence of ice the dark water absorbs heat from the sun.

During the past few decades the average temperatures in the Ilulissat area have fallen, which is in contrast to the general pronounced warming registered by meteorologists in the Northern Hemisphere. But it does appear that the cooling trend in parts of West Greenland is about to stop.

The albedo effect is the difference between the amount of light reflected by a surface, and the amount of light that falls on it. Here one can see some of the fantastic colours created.

The nutrient-rich icefjord - and the profusion of life

The meltwater from the glacier ice in Kangia contains large quantities of nutrients, and the turbulence caused by the constant motion of calving icebergs brings warm and nutritious water from the deeper layers up to the surface. Large quantities of plankton are produced all year round, and attract vast numbers of crustaceans, fish, marine mammals and birds. By contrast, the rocky land areas are poor in nutrients and heath is the most widespread vegetation type.

The rich marine flora and fauna of Kangia are closely connected with the ice and the stranded icebergs at the mouth of the fjord. As noted above this is due to the strong turbulence and up-welling of nutrients into the upper water layers. When the ice at the bottom of icebergs melts, the melt water rises along the sides of the icebergs because fresh water is lighter than the surrounding salt water. The upward currents create turbulence in the whole column of water, and this causes up-welling of relatively warm and nutritious water. The icebergs thus contribute to nourishment of the upper water layers and the flourishing plankton life at the surface. The high primary plankton production culminates in the summer, and provides the food foundation for the fjord's rich animal life.

Tidal effects, 2 to 3 metres between ebb and flood, ensure an effective exchange of water between the icefjord and Disko Bugt. This also promotes the production of plant plankton, zooplankton and crustaceans, which attract fish, seals and whales.

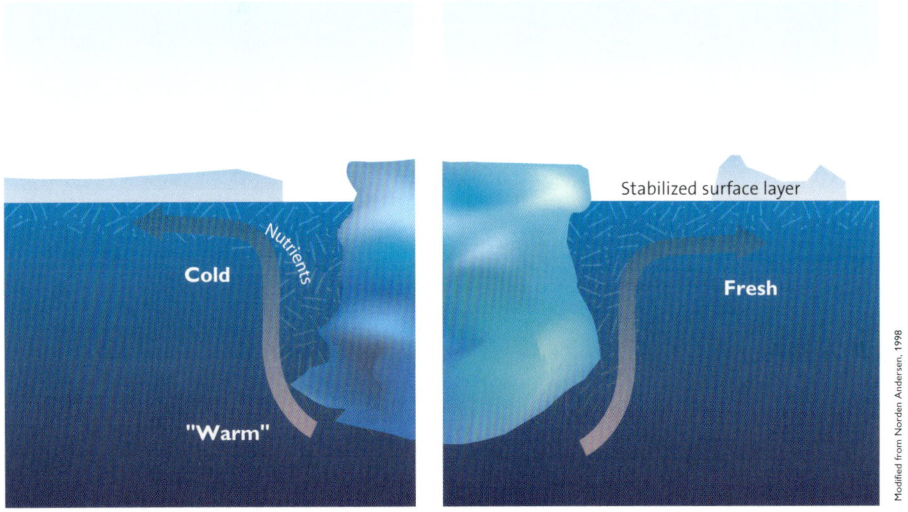

Freshwater produced by melting at the bottom of icebergs is lighter than the surrounding salt water, and moves upwards along the sides of the iceberg towards the surface. The water currents create turbulence in the water column, and bring warm nutrient-rich water up from the deeper layers, benefiting the production of plant plankton.

The tightly packed icebergs and the sea ice have hindered studies of the flora and fauna in Kangia itself. Detailed studies of life have therefore focussed on the more easily accessible areas at the mouth of the fjord.

Marine plankton

Knowledge about the occurrence and distribution of marine plankton in the icefjord is sparse. Studies in Disko Bugt show that the concentration of both nutrients and plant plankton rise markedly in the direction of Ilulissat and the mouth of Kangia, and this is without doubt due to the high primary produc-

ICEBERGS LEAD TO INCREASING AMOUNTS OF NUTRIENTS IN THE SURFACE WATER, AND TO HIGHER PRODUCTION OF ALGAE AND ZOO-PLANKTON

tion in the icefjord and the addition of nutrients from subglacial erosion. The Disko Bugt area has a very high primary production of nutrients compared with most of West Greenland. Only a very few other fjord mouths exhibit comparable high concentrations of nutrients and plankton.

Marine plankton have been investigated along a transect from Ilulissat to Qeqertarsuaq on Disko. The local processes in Ilulissat Isfjord, with melting and run-off from the glacier, can be seen to create a transport of freshened surface water that forms a vital foundation for the rich occurrences of plant plankton and zooplankton in the upper water masses.

Fish and larger crustaceans

Greenland halibut is the dominant fish species in the Kangia eco-system. Unlike other flatfish species, halibut live both at the bottom and at higher levels of the

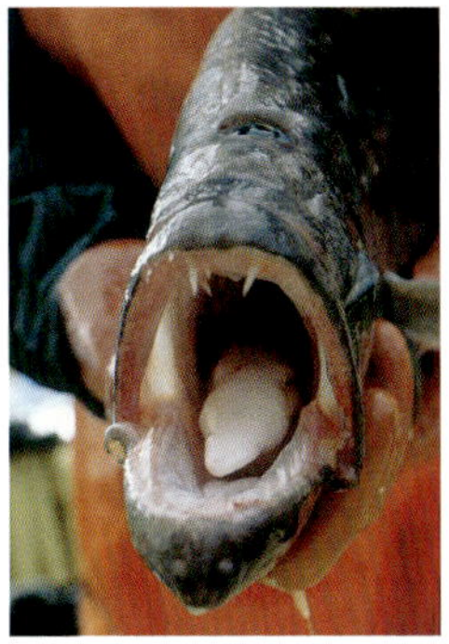

Photo: Jesper Boje

Greenland halibut is widespread in Kangia and on Isfjeldsbanken at the mouth of the fjord. The halibut can be up to 25 years old, and in rare cases become up to 120 cm long and weigh 45 kg.

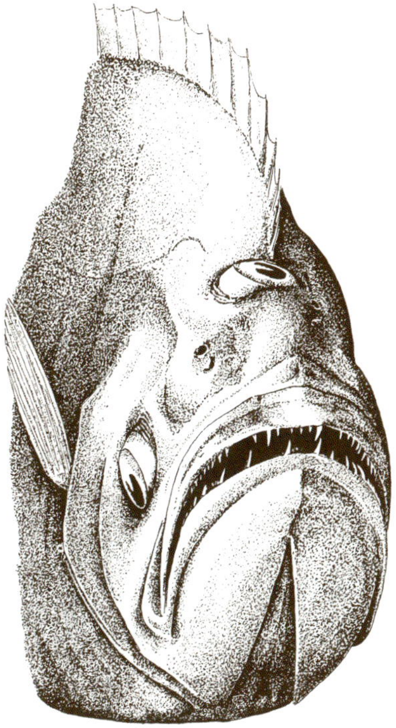

Source: Det kongl. danske vidensk. selsk.1936, Ad. S. Jensen

Illustration: © Scandinavian Fishing Year Book.

A HUMPBACK WHALE FLICKS ITS TAIL.

free water masses. Halibut flourish in cold water, in part because they are ousted by cod in warmer water. Halibut live in the whole of the fjord system, but are known for their seasonal migrations. Thus in summer halibut are mostly found in the inner parts of the fjord, while their occurrence is more evenly distributed in winter.

At the mouth of Kangia halibut are subjected to one of the most concentrated types of fishing in the world, in terms of the quantity caught in a small area. Fishing is difficult in the icefjord itself, and the fjord is thus an important 'protected' refuge area for halibut, where they can thrive in peace and become up to 25 years old.

The low bottom temperatures of the icefjord (< 2°C), suggest that halibut does not spawn in the fjord itself. The stock is therefore probably not self-recruiting, but is thought to derive from a spawning population in Davis Strait. Adult halibut in the icefjord seldom return to their spawning stock, but are believed to remain in the fjord for the whole of their adult lives. Thus, no matter how heavily the stock is fished, fishing has no impact on the spawning stock in the Davis Strait and thus on the recruitment to the icefjord. However, fishing does impact on the composition of the stock in the icefjord, and the average age of the halibut caught has fallen considerably in recent decades in line with the increase in intensity of fishing activity.

BOX 8: FISH SPECIES IN THE AREA

Fish species registered during long line fishing in Kangia:

Greenland shark
Spine-tail ray
Starry skate
Arctic skate
Sailray
Roughhead grenadier
Greenland cod
Threadfin rockling
Wolf fish
Doubleline eelpout
Greater eelpout
Vahl's eelpout
Longear eelpout
Golden redfish
Polar sculpin
Longfin snailfish
Greenland halibut
American plaice

In addition, arctic char are found in the area

Management of fishing off Kangia is thus more a matter of maintaining species diversity than of the sustainability of the halibut resource. The aim of the management is to maintain the annual catch from the stock at a level that will ensure a continued stable catch. At the same time, the authorities put weight on responsible methods of fishing that minimise any damaging effects on the bottom habitat of the fjord and the area off its mouth, through the fishing itself and through the loss of fishing tackle.

Northern shrimps are the favourite prey of halibut. Shrimps are widespread in the whole of Kangia and in the waters off the mouth of the fjord. Shrimps and other crustaceans are important prey for most fish species in West Greenland. The older shrimps seem to live on the bottom during the day, and in the higher free water masses at night.

Another important prey species is the small salmonoid fish capelin, which is found locally in large numbers, and gathers in small bays at spawning time. Farther north, capelin is replaced by polar cod, which has become more abundant in Kangia and Disko Bugt in recent years. A number of eelpout species, which live at the bottom of the icefjord, are also important prey for halibut.

New research with electronic marking of halibut shows that halibut forage in deep water as well as close to the surface. The migrations up and down through the icefjord's *c.* 1000 m deep water column occur irregularly throughout the day, and are probably connected with the presence of prey. These studies have also shown that the icefjord and the water off its mouth are integrated parts of the foraging territory of the halibut.

Halibut are themselves the prey of the Greenland shark and the ringed seal. The Greenland shark is common in the icefjord, although it has been fished intensively as a secondary catch in long line fishing for halibut. Studies in which Greenland sharks have been marked have shown that the species travels very large distances, and the sharks in the icefjord are believed to form part of the stock common to the coastal waters of the whole of West Greenland.

Marine mammals

Fin whales and minke whales are often seen in summer along the outer margin of Isfjeldsbanken at the mouth of the fjord. Humpback whales are rare summer visitors, and blue whales and Greenland right whales are rare guests. Beluga and narwhal are found in Disko Bugt in the autumn and winter, and have occasionally been seen in Kangia.

It is said locally that halibut disappear from the fishing catch totally when the beluga arrive. The halibut apparently burrow into the sediment at the bottom of the fjord to avoid the beluga, which prefer to eat halibut. The harp seal is common in open water in summer and the ringed seal lives in the icefjord all

Ringed seals live in the icefjord all the year round.

Photo: Christian Glahder

MOULTING PTARMIGAN

year around. The seals feed on fish such as capelin, redfish and polar cod, supplemented by squid and crustaceans. Polar bears, which hunt seals along the coast, are extremely rare visitors to the area.

Birds at the icefjord

The bird life at Kangia is generally typical for fjord landscapes in central West Greenland. However, large flocks of northern fulmars and gulls are particularly common around the stranded icebergs at the mouth of the fjord. Bird life on land is sparse and is dominated by snow buntings, Lapland buntings and ravens.

Little has been published about the bird life around Kangia and the following passages are based largely on information from local people and from biologists who have passed through the area in connection with other field activities.

Sea birds: There are a number of sea bird breeding colonies along the coasts in the Ilulissat area, including colonies of Iceland gull and kittiwake, and a few of cormorants. The most impressive sea birds are the large flocks of northern fulmars, which find their food amongst the grounded icebergs on Isfjeldsbanken. The fulmars come from large breeding colonies on Disko and on the small island Qeqertaq in Disko Fjord, about 100 km west of Ilulissat. The population is thought to number more than 100,000 breeding pairs. Large flocks of Iceland gulls and glaucous gulls forage frequently amongst the stranded icebergs, together with lesser numbers of great black-backed gulls and kittiwakes. The fulmars and the gulls are attracted by the zooplankton in the waters at the mouth of Kangia, as well as discarded fish refuse from the trawlers and boats fishing for halibut.

Red-throated diver on its nest.

Land birds: The relatively sparse bird life on land includes the widespread Greenland species such as snow bunting, Lapland bunting, redpoll, wheatear and raven. The Greenland white-fronted goose is also found. Geese fitted with small transmitters were traced to the area by satellite in the spring of 1999, and according to local hunters breed on Nunatarsuaq. The breeding area for Greenland white-fronted goose is limited to West Greenland, and is estimated to comprise a total of only 30,000 individuals, which all winter in the British Isles. Other Greenland birds found in the Ilulissat area include peregrine falcon, gyr falcon and ptarmigan, all of which are believed to breed here, at least occasionally. The little arctic redpoll appears as a winter guest from more northerly areas in Greenland. The red-throated diver, the great northern diver, Canada goose, mallard, long-tailed duck, red-breasted merganser, red-necked phalarope and purple sandpiper also probably breed in the area.

Visiting birds: A large number of bird species are found as more or less regular guests in the area, either on migrations in the spring and autumn or as summer guests. These include, for example, brent

Colony of great cormorants and Iceland gulls

Photos: David Boertmann

Gulls feeding on fish refuse in the icefjord

Location of sea-bird colonies in the Ilulissat area.

Colony number 1 is a great cormorant (Phalacrocorax carbo) and Iceland gull (Larus glaucoides) colony. In 1976 there were 170 pairs of Iceland gulls, while the number of cormorant nests was unknown. Cormorants were noted at colony number 2 in 1995, but breeding was not confirmed. Colony number 3 is a kittiwake (Risa tridactyla) colony, with in 1976 a total of 15 pairs in company with 100 pairs of Iceland gulls. The two cliff colonies, numbers 4 and 5, are probably Iceland gulls. Colony number 6 is probably a breeding colony of Iceland gulls. Colony number 7 on a steep cliff also comprises Iceland gulls, and further colonies of gulls have been reported at sites 8 and 9.

REINDEER ON THE RUN

goose, harlequin duck, eider, pomarine skua, arctic skua, arctic tern and Brünnich's guillemot.

Land mammals

Available information about mammals on land in the Ilulissat area is limited, but the arctic fox is quite common. The arctic fox is one of the warm-blooded animals that have adapted best to the cold. In winter they can keep warm in temperatures as low as minus 40°C without moving, due to their highly insulating fur. The arctic fox eats anything and everything, and has an amazing ability to adapt. It catches fish in the tidal zone, takes birds, eggs and insects, feeds on carrion and it also eats abundant berries in the autumn.

The arctic hare is fairly rare in the area and is mainly found in the higher-lying areas near the margin of the Inland Ice. It eats a broad range of plants, and survives the arctic winter by burrowing its way down to plants in areas with little snow cover.

Adult arctic hares can weigh up to five kilograms, more than a fully grown fox, which does not usually hunt them. Arctic hares are often seen in flocks in winter.

Reindeer are found locally in the area south of Kangia. However, they have not been recorded from the areas on the north side of the fjord in modern time. Some place names indicate that they have lived there in the past, and this is supported by finds of reindeer bones from the Dorset and Thule cultures in the midden at Sermermiut. Reindeer are the only mammals that regularly eat lichens. In winter the reindeer reach the lichen by scraping through the snow with their hooves.

Plants on rocky ground

The flora in the Ilulissat area is low arctic and reflects the nutrient-poor, silica-rich rocky ground. About 160 species of flowering plants, three species of club-mosses, two species of horsetails and four species of ferns are known from the area, and the plants colonise new areas as the ice margin retreats. Some species are seen nearly everywhere, while others are only found at a few locations. How-

The arctic fox eats anything and everything, and has an amazing ability to adapt.

ever, even the most common species have their preferred growing sites in one or several characteristic plant communities.

The dominant plant communities are heath, fell-field, snow-patch, herb slope, willow-scrub, fen, riverbank and seashore vegetation. In addition there are a number of plants that grow in lakes and ponds.

Heath: As in many other parts of Greenland, the most widespread plant community is heath, dominated by dwarf-shrubs which form a uniform growth over large areas. Dwarf birch, crowberry and arctic blueberry are the typical species, of which the last two give a good yield of berries in the late summer. Arctic bell-heather is common on north-facing slopes. In the most sheltered areas, mountain heath can form dense growth with thousands of red-lilac, bell-shaped flowers. Other common heath plants are the strongly aromatic, stunted bush Labrador-tea and fir club moss, which stretch their curved, green fingers up from the moss. The little white-flowered herb, long-stalked stitchwort, is rarely absent. Grasses such as the arctic meadow grass and Alpine sweet grass often stand out among the stunted bushes. Rock cranberry is also found, but is rare.

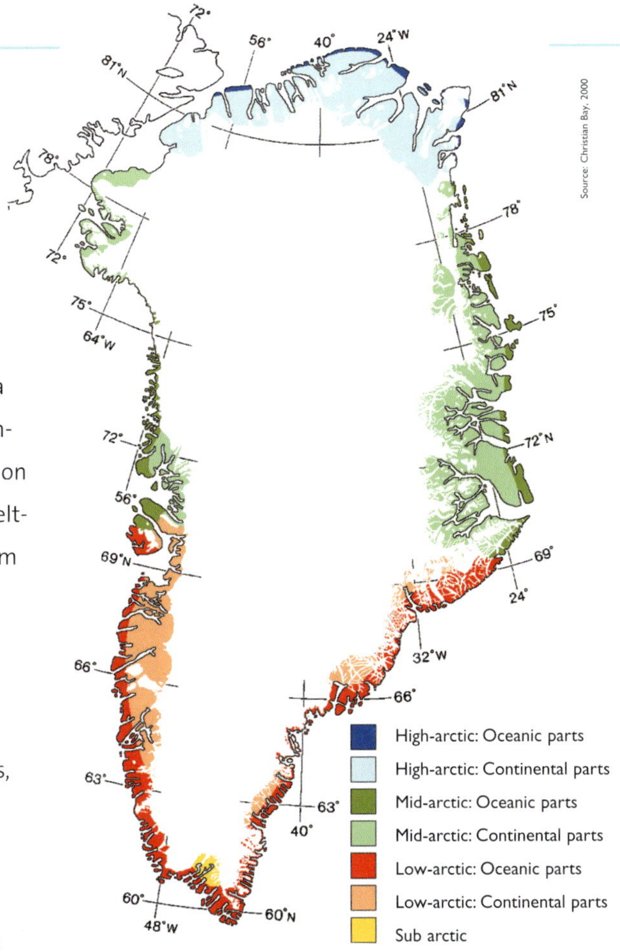

Phytogeographical zonation of Greenland.

PURPLE SAXIFRAGE

The bushes are more widely spaced where the heath is subject to strong drying winds. This gives room for herbs such as the beautiful hairy lousewort, which stands out with its light red inflorescence and the white tomentose Labrador cat's-paw. Lapland rose-bay, with leathery leaves and large blue-lilac flowers, is found on the more nutritious places, and fits in well with entire-leafed mountain avens, which also grow here with their large, whitish-yellow flowers.

Most ferns avoid dry conditions such as sunny rock crevices, although the fragrant shield-fern does grow at such sites. Brown, withered leaves several years old protect the new dark-green leaves. Three-toothed saxifrage, which is one of the few prickly plants in Greenland, also prefers dry rock cracks.

Fell-field: Plants are more scattered on the fell-fields. The distance from one green tussock to the next can be long, either because it is too dry or too windy, or because the soil is so coarse-grained that it cannot retain moisture. Nevertheless, flowers can nearly always be found if one looks for them, and there are often many species, because the competition between species is minimal. Snow whitlow-grass and Lapland diapensia are common, both with white flowers, in addition to the yellow-flowered arctic poppy and the butter-yellow snow cinquefoil with white undersides to its leaves. The

Arctic poppy

small, dark head of the northern wood-rush is also frequent on the fell-fields.

Patterned ground and other interesting soil phenomena caused by frost action are common. Polygonal patterns are developed in stony soil, with the coarser grained material along the edges of the polygons and finer material in the cen-

White Arctic bell-heather

tres. During the thaw, the centres of the polygons are soft and sticky because the melt water cannot seep down into the permanently frozen soil. Later in the summer the upper soil dries out. Plants that can survive on this type of 'living' soil need special abilities. For example, grained saxifrage forms scores of small leaf buds instead of flowers, which

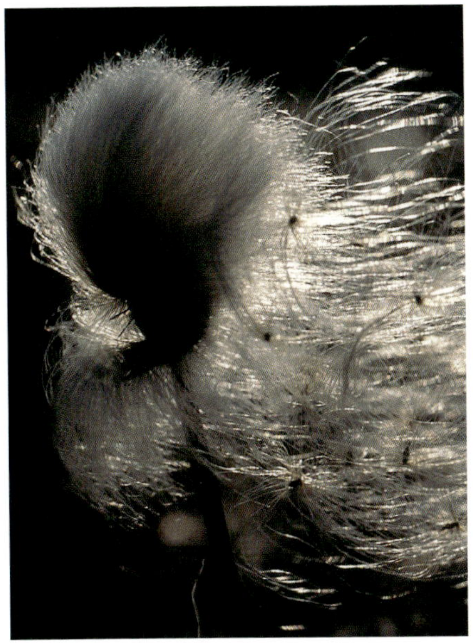

Cottongrass

Mountain sorrel

Willow

detach from the inflorescence and sprout in the soft soil.

Snow-patch: Places where the snow remains for most of the summer are called snow-patches. Here the growing season for plants is only four to six weeks long. For the rest of the year the plants are covered by a thick, protective layer of snow. Mosses, lichens and dwarf willow are found in snow-patches. The branches of dwarf willow are often hidden in the soil, with only the leaves and inflorescence protruding above the surface. The male plants have yellow inflorescence, while those of the female plant are red. Moss heather is found in the more heath-like snow-patches. Other characteristic herbs are the annual pigmy buttercup, with small, yellow flowers, dandelion and mountain sorrel. It is possible to make a nourishing fruit soup using mountain sorrel:

by boiling a large bunch of stems in a pot until tender, straining the juice and adding sugar to taste.

Herb-slopes: The most lush plant community in the area is the herb-slope, or 'urteli', which carries a vegetation very rich in species on strongly sloping areas. 'Urte' means herb in Danish and Norwegian, and 'Li' is a 'slope' in Norwegian. Herb-slopes develop on the south or south-west facing slopes beneath high, steep mountains, where they receive meltwater throughout the summer and receive the maximum amount of light from the sun; this creates a warm and mild microclimate. At the same time they are fully protected in winter by a thick layer of snow. Thirty different species of flowering plants can grow on such slopes, including species such as Alpine bistort, Unalaska fleabane and Alpine bartsia, with dark-red flowers and bracteoles. Between the higher herbs one can be lucky to find the little thick-leafed whitlow-grass with its light-yellow flowers.

BROAD-LEAVED WILLOW HERB

Seashore: Seashores have their own vegetation, with sandy shores having the characteristic species sea sandwort and lyme-grass. Gravel sedge, sea plantain, Greenland scurvy grass and low stitch-wort are found on coasts with rocks and gravel. On clayey coasts grassy veget-ation develops, that is dominated by creeping saltmarsh-grass, with Pacific silverweed with its yellow flowers bright-ening the upper part of the shore.

Hairy Lousewort.

Willow scrub: Willow scrub around Kan-gia can grow to considerable heights in the most lush valleys, in some cases up to 1.5 m. In most places the willow scrub is only knee high, but for the plants at the bottom of the scrub this is of little importance as long as they have shade and shelter. Here one can find interrup-ted club-moss and common horsetail and, in the drier scrub, round-leafed wintergreen with large flowers.

Stony riversides: Stony riversides are widespread in the area. Here seasonal variations in the water supply result in frequent flooding. In this unstable envi-ronment the most conspicuous species is the broad-leafed willow-herb with its pink flowers. This is Greenland's national flower, and the poetic Greenlandic name 'niviarsiaq' translates as 'young girl'.

Fen: The richest fens and mires have a thick vegetation of grass-like plants, often dominated by arctic water sedge with its thick stem and upright female spikes, and loose-flowered Alpine sedge which carries beautiful hanging female spikes. Arctic marsh willow, which has branches that creep through the veget-ation, is rarely absent; the beautiful red catkins can be up to 10 cm long. The yel-low 'candlesticks' of flame-tipped louse-wort are also common, while the frail Lapland buttercup is rarer.

Lakes and ponds: Mares-tail is the most common of the proper aquatic plants. The characteristic green 'Christmas trees' stand half-covered by water along the banks of many ponds, small lakes and slowly flowing rivers. Northern bur-reed has long ribbon-like leaves that float on the surface of the water and are aligned by the prevailing wind; only the infloresc-ence is above the water-surface. Comple-tely covered by water is the small pondweed, white water-buttercup and, here and there, a very small green leaf rosette of water awlwort, which only flowers if the pond dries out. Freely floating in the water may be found the quite rare autumnal water-starwort. In many places there is marsh-like vegetation on the banks of lakes and ponds, where one can see plants such as bleaked sedge and arctic cotton-grass. Where water trickles throughout the summer the insect-eating butterwort grows, whose sticky, enzyme-rich leaves hold and decompose small animals. The little mudwort thrives in rock puddles, even if the puddle dries up.

Entire-leafed mountain aven.

Traditional use of plants

Although the Inuit are usually considered strictly meat eaters, plants have traditionally formed a part of their diet. Mountain sorrel, crowberry and arctic blueberry are among the plants that continue to be gathered and eaten. Their nutritious contribution to the diet is no longer as important as it once was, because fresh fruit and vegetables are flown in from Denmark to even the most remote settlements. However, the local people still consider gathering and eating traditional Greenland food as an important part of their culture.

The use of plants for fuel and handicrafts was common until recently. Because of their oil-containing shoots, white arctic bell-heather and crowberry have been used to a large extent as fuel for outdoor use. Even when freshly picked they burn well. Families still use crowberry as fuel when they go hunting in the fells. They find a suitable space between the stones, make a fire with crowberry and cover the fireplace with a flat stone on which the meat is grilled. Crowberry is also used south of Kangia for quickly smoking arctic char.

Lichens

SOIL WITH LICHENS

LICHENS GROWING ON THE GROUND

Flavocetraria nivalis is among the most characteristic lichens occurring in heavily wind-blown fell-fields. It usually covers large areas with its bright, yellowish white thallus.

LICHENS GROWING ON THE GROUND

Photos: Eric Steen Hansen

Nephroma arcticum has broad, yellowish green lobes, which grow among the dwarf shrubs in moist heaths.

The mainland around Disko Bugt is one of the best-investigated areas of West Greenland in respect of lichen studies. About 200 species are known from the Ilulissat area. It is rich in species that grow on the ground and on rocks, while there are only very few lichens living on bark. Some lichens such as *Cladonia stellaris, Lobaria scrobiculata, Nephroma arcticum* and *Physconia detersa* have their most northerly occurrences in the area. A few high arctic species such as *Arctocetraria nigricascens* have their southernmost outposts here. Information about lichens can be found in the book 'Greenland Lichens' by E.S. Hansen (1995).

Lichens growing on the ground

Most of the heath types in the area are rich in lichens. Humid heaths have species such as *Alectoria nigricans, A. ochroleuca, Bryocaulon divergens, Bryoria nitidula, Cetraria islandica, Cladonia mitis, C. stygia, Flavocetraria cucullata, F. nivalis, Peltigera scabrosa, Sphaerophorus globosus, Stereocaulon alpinum, S. paschale* and *Thamnolia vermicularis.*

In marsh-like heaths, *Flavocetraria cucullata, Icmadophila ericetorum, Omphalina hudsoniana and Thamnolia vermicularis* frequently grow on the tops of tussocks. Bare heath soil is occasionally colonised by *Arthror-haphis citrinella, Candelariella placodizans* or *Lecanora geophila.* In areas with fell fields with scattered crowberry and arctic blueberry, one will also find *Cetraria nigricans, C. muricata, Cladonia borealis, C. pyxidata* and *Ochrolechia lapuënsis. Alectoria ochroleuca, Caloplaca jungermanniae, C. tiroliensis, Flavocetraria nivalis* and *Physconia muscigena* in particular grow in drier heath communities dominated by sedges.

Meltwater channels that cut through the heaths and fell fields are the preferred growing place for *Arctocetraria andrejevii, A. nigricascens, Cetrariella delisei* and *Cladoniia trassii.* Heaths with Lapland rose-bay are characterised by the occurrence of *Alectoria ochroleuca, Bryoria nitidula, Cladonia mitis, C. stygia, Flavocetraria cucullata, F. nivalis, Ochrolechia frigida* and *Sphaerophorus globosus.*

The soil of snow-patches often have quite low acidity, where species such as *Cladonia ecmocyna, Lepraria frigida, Peltigera malacea* and *Solorina crocea* flourish.

Rock-growing lichens

Lichen vegetation growth on rocks and boulders near the icefjord is dependant on temperature, humidity, the mineral composition of the rocks, the slope of the rock surfaces, their exposure to the elements, and notably the presence of the excrements of birds and animals. Gneissic rocks lacking additional fertilisation sources are often covered by dark lichens such as *Cetraria nigricans, Melanelia hepatizon, Pseudephebe minuscula, Orphniospora moriopsis, Umbilicaria hyperborea* and *U. torrefacta.*

ROCK-GROWING LICHENS

Photos: Eric Steen Hansen

Umbilicaria decussata has a greyish, foliose thallus with distinct reticulate ridges. It grows here together with Rhizocarpon geographicum and different black micro-lichens manured by birds.

ROCK-GROWING LICHENS

The orange-red species, Xanthoria elegans, binds gravel with its thallus and in this way prepares the substrate for later colonisation by higher plants.

ROCK-GROWING LICHENS

Pseudephebe minuscula forms concentric rings on the rock face. The central part of the thallus dies allowing new thalli to start their growth inside the outermost ring, which often is more than 100 years old.

Rhizocarpon geographicum and *R. inarense* form a contrast to the dark lichens, with which they compete for room. These communities are best developed at some distance from the Inland Ice. *Pseudephebe minuscula, Rhizocarpon geographicum* and *Umbilicaria hyperborea* start to colonise the rock surfaces just a few years after the ice has disappeared. Minute lichens are found close to the margin of the Inland Ice, and they become progressively larger with increasing distance from the ice margin. Other species such as *Brodoa oroarctica, Calvitimela armeniaca, Hypogymnia austerodes, Ophioparma ventosa, Parmelia saxatilis* and *Sphaerophorus fragilis* are often found growing on silicate-rich rocks.

Bird-stones are rocks and boulders where birds regularly perch, and their droppings provide a source of nutrients. These stones are often strongly coloured by *Physcia caesia, P. dubia, Rhizoplaca melanophthalma, Umbilicaria arctica, U. decussata, Xanthoria borealis* and *X. elegans*. Some wide rock crevices near the icefjord are of particular interest because of the occurrence of the two species *Lobaria scrobiculata* and *Peltigera collina*, which are characteristic of areas with a continental climate. On the vertical and east-facing rock walls of the icefjord these two lichen types grow on top of mosses together with *Lecanora swartzii, Parmelia saxatilis, P. sulcata, Physconia detersa* and *Xanthoria borealis*. Another wide ravine in the area, where raven droppings are abundant, is rich in lichens such as *Acarospora peliocypha, Amandinea cacuminum, Lecanora argopholis, Phaeophyscia endococcina, P. sciastra, Rhizocarpon geminatum* and *R. grande*. Rocks close to the coast that are influenced by gull excrement are often coloured yellow by *Candelariella arctica* and *Lecanora straminea*. Rock surfaces at the water's edge are often covered by the black thallus of *Verrucaria ceuthocarpa*.

Rusty weathering iron-rich rocks often have rust-coloured (iron tolerant) lichens such as *Acarospora sinopica, A. smaragdula, Lecidea silacea, Miriquidica atrofulva* and *Tremolecia atrata*. As a rule, the lichens grow under quite humid conditions on stones in marshes and near streams.

Lichens living on bark

Bark-dwelling lichens that grow on bushes and dwarf-shrubs are rare as far north as Ilulissat. The following species have been found on the bark of northern willow in the area: *Caloplaca borealis, C. cerina, Caloplaca tiroliensis, Candelariella xanthostigma, Lecanora fuscescens* and *L. symmicta*. Bark-growing lichens grow only in warm places protected against strong winds.

The land of the 'icefjord people'

THE COASTAL CLIFF AT SERMERMIUT

Many archaeological finds, and saga narratives and legends in more recent years, testify to the long history of settlement in the harsh wilderness of the Ilulissat region, where hunting and fishing have always been the basis of life. In modern Ilulissat, fishing and the fishing industry continue to be the most important business enterprise in what is Greenland's third-largest town.

Traditional winter house from the Thule culture, modified by a glass window. The Thule people, ancestors of today's Inuit, settled at Kangia 800 years ago.

Cultural changes amongst the people living near Bering Strait, between Siberia and Alaska, about 6000-5000 years ago appear to have led to the development of a new culture adapted to the arctic environment, and this culture spread eastwards. This Palaeo-Eskimo group is known internationally as 'the Arctic Small Tool Tradition'. The first representatives of this culture reached West Greenland 4500 years ago and are known as the Saqqaq people.

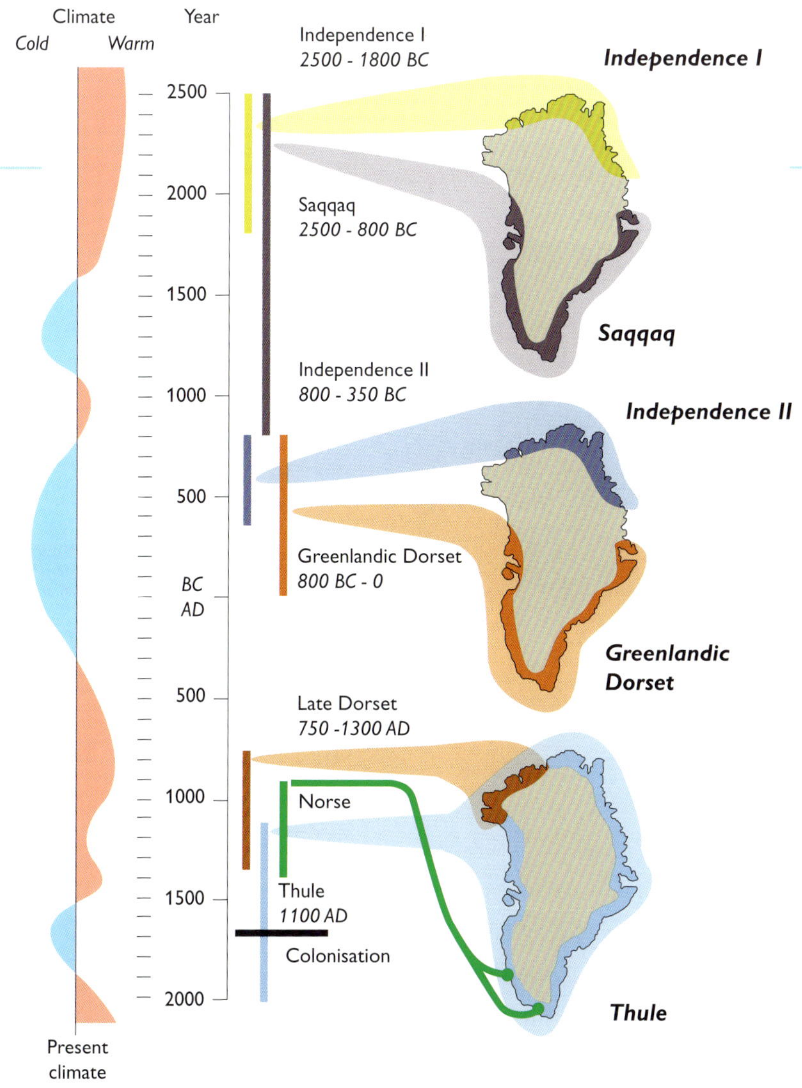

Chronology of the different cultures that have colonised Greenland, and a generalised climate curve for the last 4500 years.

Modified from Hans Christian Gulløv

The Saqqaq people settled in the Ilulissat area about 4400 years ago. They lived around the icefjord Kangia for 1500 years, and a little longer at other places, before they disappeared. Then there was a gap in the human settlement of West Greenland until the arrival of the Dorset people in the area about 2700 years ago. After 700 years of settlement, the Dorset culture in its turn disappeared, and the Disko Bugt region was uninhabited for a thousand years until the people of the Thule culture settled at Sermermiut and Qajaa 800 years ago.

Palaeo-Eskimo settlements

Palaeo-Eskimos, equivalent to the Stone Age cultures of Europe, are represented by four cultures in Greenland. The Saqqaq and Dorset people lived in West Greenland, while the Independence I and Independence II people settled in the northern and north-eastern parts of Greenland. All of the palaeo-eskimo cultures migrated eastwards across Canada, reaching North-West Greenland in the Thule area. From here the Saqqaq and the Dorset people moved down the west coast and established settlements at suitable locations.

Palaeo-Eskimos lived by hunting and fishing. Archaeologists have investigated a large number of settlements, especially along the coast south of Ilulissat, but none of them are as large or as important as Sermermiut and Qajaa near Kangia. These large settlements, and the dense pattern of smaller settlements, indicate that the resources available around the icefjord were both rich and stable.

For hunting, the Palaeo-Eskimos used lightweight harpoons, sometimes fitted with a tip of stone, as well as lances, knives and bows; bows and arrows are not known from the Dorset period. At the settlements, tools such as scrapers, augers, knives and needles of bone were

SEAL BONES FROM A MIDDEN AT SERMERMIUT

used, while knives and burins were used both at home and outside. Weapon tips and cutting tools were made of silicified slate in the Saqqaq period and of chalcedony in the Dorset period. Most of the tools had wooden shafts, but in some cases the shaft was made of bone with lashings of sinew or baleen.

The Saqqaq probably used kayaks, while sledges drawn by the hunter himself or by a few dogs are known first from the Dorset culture. Their homes were skin tents used all year long, which were anchored to the ground using stones or poles. In the centre of the tent was a fireplace, sometimes placed in the middle of a rectangular stone framework with room for the fire and firewood, and which divided the tent into two equally large parts. The Saqqaq people used small round lamps of soapstone, while those of the Dorset people were oval.

At the midden at Sermermiut, bones and tools made of wood and bone in the lower layers were poorly preserved. The Qajaa midden is in the permafrost, and preservation conditions are unusually good. The bone remains show that the Palaeo-Eskimos made good use of the resources of the area. Bone remains of ringed and harp seals dominate, but they also hunted arctic fox, polar bear, arctic hare, walrus, narwhale, beluga and sea birds.

Driftwood was a very important resource, and was a necessity for making tents, kayaks, sledges, bows and lances. The ocean currents provide a continuous supply of driftwood, derived from Canadian and Siberian forests.

Thule people at the icefjord

The Thule people arrived in Greenland about 1100 AD, and settled at Sermermiut 100 years later. Their winter houses were built of stone and peat with a roof construction of driftwood or whale ribs. During the summer they lived in tents. The Thule people exploited the same animals as earlier cultures, but they also hunted large whales and fished with long lines and nets made of baleen.

Photo: Erik Holm

A 3500 years old burin from the Saqqaq culture, found at Qajaa. The flint blade is attached to the wooden handle with baleen. Burins were used for splitting bone and wood.

The Thule people were much more mobile than their predecessors thanks to their sophisticated kayaks, their dog sledges and the large skin boat, the umiaq, which was used during whale hunting, or for transporting the whole camp from place to place.

Finds from Sermermiut and Qajaa indicate that the Thule people had contact with the Norse, who came from Iceland and northern Europe and who lived in South-West and western Greenland from about 985 AD. The Inuit probably obtained iron partly from the Norse, and partly

Photo: Hans Kapel

View of the Sermermiut valley, with Isfjeldsbanken in the background.

from meteorites in the area south of Thule. The Norse are thought to have disappeared from Greenland around 1450 AD, and their fate is unknown.

In the 16th century, European explorers visited Greenland waters, followed by whalers in the 17th and 18th centuries. Both groups had contact with the local Inuit, and around Disko Bugt there were several meeting places where trade was

Source: Greenland National museum in Nuuk

Map of the Ilulissat region, with circles marking former eskimo settlements. The largest and most important settlements were Sermermiut (14) on the northern side of the icefjord and Qajaa (2=Qajâ) on the south side about 25 km east of the mouth. The various Thule settlements around Kangia, including Sermermiut and Qajaa, date from about 1200. The map uses many, now obsolete Danish place names.

Photo: Jørgen Meldgaard

ARCHAEOLOGICAL EXCAVATION OF A MIDDEN AT QAJAA. LAYERS OF THE MIDDEN DATE FROM THE SAQQAQ, DORSET AND THULE CULTURES

carried out. A few settlements along Greenland's west coast became centres with quite large populations, and that at Sermermiut was the largest of them all.

Sermermiut in Greenlandic means 'the place of the glacier people', and it was a good place. The highly productive and stable eco-system of the icefjord formed the foundation for the success of Sermermiut and Qajaa for thousands of years. With its location at the central part of the icefjord, Qajaa was vulnerable to the advances of the glacier front, and the settlements here coincide with periods when the glacier had retreated, and living conditions were most favourable for human existence.

The long but sporadic contact with Europeans became more regular after the Danish colonisation, which started in 1721 with the foundation of Godthåb, the present day Nuuk. Jakobshavn, now Ilulissat, was established as a colony in 1741 a few kilometres north of Sermermiut. In the course of the next century, Sermermiut was gradually abandoned by its inhabitants, most of whom moved to the new town; the last inhabitants left about 1850, and many aspects of their former lifestyle vanished with them.

Rich archaeological finds

Archaeological excavations at Sermermiut and Qajaa have played a major role in our understanding of the history of human settlement in West Greenland.

Sermermiut is exceptional because the site had been inhabited during all three of the cultural phases of West Greenland. The former inhabitants of Sermermiut were very much aware that they lived in a special place. In 1737, the Danish missionary Poul Egede wrote in his diary:

Source: Modified from Tina Møbjerg, 1983

Map of the Sermermiut valley with ruin sites indicated, mostly winter houses from the Thule culture.

Visitors to the sensitive and valuable Sermermiut site are guided by information posters.

"I found here the largest group of people I have seen anywhere in Greenland, with about 20 very large houses, like a little village. They boasted about this and asked if I had seen so many people at one place elsewhere. I felt immediately from their speech and manners that they were proud of their number and the good catches they had."

The middens at Sermermiut and Qajaa contain an outstanding sequence of three cultural layers separated by sterile peat layers. The bottom layer, from the Saqqaq people, is clearly separated from the middle Dorset culture layer. A new sterile layer follows, and is succeeded by a layer with remains from the Thule culture. At Sermermiut the stratigraphy is visible in a low coastal cliff that is slowly being eroded by the sea.

Archaeological excavations at Sermermiut started after the Second World War. At that time it was believed that Greenland's pre-history was only about 1000 years old. The excavations at Sermermiut showed without doubt that Greenland's cultural history was much longer and more complex than previously thought, a conclusion supported by contemporary investigations in North Greenland. The oldest layer at Sermermiut was dated to be almost 4000 years old. Later excavations at Qajaa confirmed the existence of three different cultures in West Greenland.

The pre-historic cultures at Qajaa were actually discovered as early as 1871, when a Greenlander, Carl Fleisher from Ilulissat, made the first excavations there. The same summer he reported the presence of several midden layers, and of his finds from the bottom of the midden wrote that they represented a time when 'the old Greenlanders used stones for their tools'. Carl Fleisher was the first to formulate the idea that another cultural lifestyle had preceded the traditional lifestyle of his day. Unfortunately, no-one had knowledge of his important finds and interpretations until his letters were rediscovered more than 100 years later. Qajaa was abandoned at the beginning

Photo: Jørgen Meldgaard

Excavations at Qajaa, seen from the leader's tent where the finds are registered and drawn.

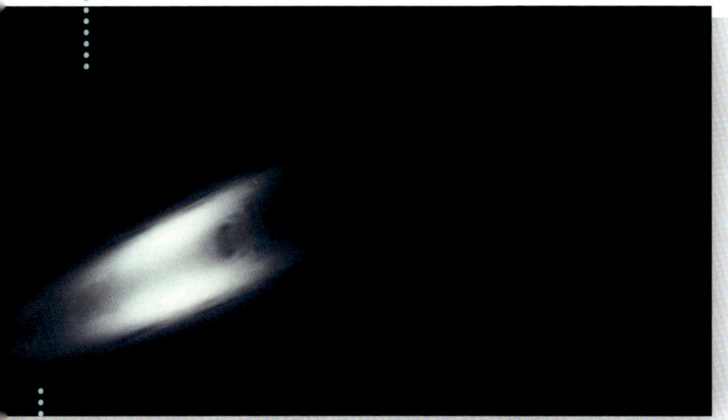

ANNINGAT (THE MOON) REFLECTED IN THE BLACK WATER

of the 18th century, while Sermermiut was inhabited until about 1850. The ruins of the turf houses that can still be seen at Sermermiut reflect the latest period. From an archaeological point of view these houses are especially valuable, because finds from excavations can be compared with information from written Danish sources and the oral traditions of the Inuit. Both Sermermiut and Qajaa still offer great potential for studies of the pre-history of Greenland.

Eqi - a settlement at the mouth of the icefjord

There are several smaller settlements around the icefjord that have not yet been thoroughly excavated. These sites are often only known from surface finds of tools that give an indication that the Saqqaq or the Dorset people have been at the place. The smaller settlements are found in particular on the south side of Kangia near its mouth, and down along the coast to the far side of the village of Ilimanaq.

One of these sites is Eqi, which lies on the south side of Kangia's mouth opposite Sermermiut. Eqi is a windswept place with few possibilities for shelter. However, in sunny calm weather the place has a harsh beauty of its own, with polished granite rocks, knee-high green grass and lyme-grass forming a contrast to the blue water and the chalk-white icebergs that tower over the coast. The location was very well situated for hunting and fishing in the icefjord.

The deserted settlement of Eqi lies on a small ledge beside the fjord, with its back to the undulating hinterland. One can still see the collapsed ruins of the last turf houses, which resemble low grassy hillocks. The whole of the former settlement area is one enormous midden that dates from the period from 4000 years ago up to about 1950.

In 1933, the archaeologist Therkel Mathiassen visited Eqi. He recorded several ruins of winter houses from which material had been taken to construct later buildings, and he also found bone remains of whales and other animals. In 1953, the archaeologists Helge Larsen and Jørgen Meldgaard dug a quick test trench in which they, not unexpectedly, found flint refuse but no tools. Close to the trench, in the tidal zone where the

The oral tradition in Greenland is rich and flourishing and can often provide more information than scientific investigations based on written sources and archaeological finds.

For example, there is a story of the first man to settle at Sermermiut. His name was Qingernilik, and he and his family experienced a harsh first winter. Suffering from hunger, they crossed the icefjord to reach the village of Eqi, where there was plenty of food, and Qingernilik ate so much that he almost died. Qingernilik returned to Sermermiut, but he never visited Eqi again because he did not want to take the risk of dying through over-eating.

Another important person in the history of the icefjord is the great angakkoq, or Eskimo shaman, who taught the people where they could fish for halibut. It is said that when Inuit people from Sermermiut were on their way home from Eqi on the south side of the icefjord, the angakkoq suddenly collapsed on the ice with a loud crash. He told the people travelling with him that they should fish at this point. The people made a line of baleen, hacked a hole in the ice and started to fish, and immediately caught some good halibut. Thus began the fishing tradition that has since been a source of wealth and survival for the population.

Myths and religion

Everyone who lived in the area was aware that Kangia was important to the whole world, because it was here that the sun and moon originated.

BOX 10: FAITH AND STORY-TELLING TRADITION

Illustration: Naja Abelsen, Scene from art-cartoon Solen og Månen, Milik Publishing, Nuuk.

The story of their creation is the story of a brother, Anningat, and a sister, Ajut, who before they became the moon and the sun lived at the icefjord. Their transformation arose because Anningat lusted after his beautiful sister, and he continued to hunt Ajut until they were both sucked up into the air where he, the moon, still hunts her, the sun.

This story is a part of Inuit mythology and religion. Although the sun is the driving force in the world, the moon is the dominant person in the myths. He is more changeable than any other heavenly body as he grows, becomes round and shrinks before disappearing for a period. In West Greenland it was believed that when the moon disappeared he was down on the ice-covered sea hunting seals. That the moon brings about the changes in the tide and influences women's menstruation was an observation that was common knowledge all over Greenland.

The Inuit believed that the world was divided into two - a visible world, which everyone could see with the naked eye, and an invisible world, which could only be seen by people who had an inner eye. The angakkoqs, the shamans, could use their inner eye to replace normal vision so they could see what was hidden, although they could not see both worlds clearly at the same time. However, in a trance-like state induced by traditional drum ceremonies the angakkoq could direct his inner eye at another person and see that person's hidden world and inner circumstances, which could be in trouble and require treatment.

The most skilled angakkoqs could travel in the invisible world. The two worlds were mutually connected, and it was the life in the other world that put limitations on the visible world, manifested through the movements of the heavenly bodies and the changes of the seasons.

All the Inuit in Greenland had more or less the same understanding of their environment. When people from near and far gathered, they would entertain each other with stories and participate in the same shamanatic seances without problems of understanding.

The old stories

The Inuit differentiate clearly between newer and older stories. The newer stories dealt with events that they had second-hand or third-hand knowledge about from the person who had experienced the events. On the other hand, the older stories (or legends) were handed down from their forefathers through many generations and - unlike the newer stories - had to be told without any changes.

That the stories nevertheless developed is inevitable in a verbal tradition. Different versions of the same basic story were written down in the colonial age. Every storyteller had his own personal style, but it was vital that the storyteller told his own version as he or she had understood it. The requirement was not so much a word-for-word rendition but a reproduction with unconditional respect for the original.

The idea was that the legends contained the wisdom of their forefathers, which should not be lost under any circumstances. The oldest of these forefathers, who lived when the world acquired its final form, were considered to be stronger and cleverer than the people of today. They had contributed to the world becoming what it is. It was these respected forefathers who had discovered the secret of how people should live with each other and with their prey, so that the Earth could continue to exist and life continue for their children, and their children's children, for ever.

Photo: Joel Berglund

RUIN OF WINTER HOUSE AT SERMERMIUT

sea washes the soil away, the archaeologists found the remains of a fireplace and numerous tools of slate and a flint-like material.

By far the majority of the tools found are from the Dorset period. They include knives, burins and micro-flakes. Only two items dating from the Saqqaq culture were recovered, a burin and a narrow knife blade.

During a visit to Eqi in 2003, it was still possible to collect flint refuse in the tidal zone, and there is no doubt that the site still has valuable archaeological potential. The locality was probably chosen primarily on the basis of the good hunting possibilities, but one cannot ignore the possibility that the majestic view and the fascination of the ice played a role in the selection of this site.

Vivid first-hand descriptions of contacts with the people of Eqi are available to us from Poul Egede, who described several visits to the settlement during the 1730s in his diary. The descendants of the Thule people are given form and voice, as the missionary discusses theology with the settlement's heathen shaman. The people talk of their hunger, and of dogs that run about with the bones of the dead, due to careless internment of people without family to bury them properly. During another visit to Eqi, Poul Egede hears about a woman who mysteriously disappeared. Apparently

BOX 11: HALIBUT, HUNGER AND SURVIVAL

Drawing: Jens Rosing

Long-line fishing for Greenland halibut as practised at Kangia since the beginning of the 19th century. Note the glider, and the length of the line.

Throughout history, the rich eco-system of the icefjord has made it possible for a relatively large population to live by the fjord. At the end of the 19th century over 60% of the people in the district lived within a distance of 3 km from the icefjord.

All of the Palaeo-Eskimo cultures that settled here have doubtless caught halibut, but despite numerous archaeological excavations around Disko Bugt not a single bone of the fish has been found. This may be because halibut bones have a poor preservation potential compared to other fish bones, as they are rich in fat and are more cartilage than bone.

The earliest written documentation of fishing for halibut in Kangia is from February 1739, and comes from Poul Egede's diary. When he travelled between Qasigiannguit and Sermermiut he met fishermen on the fjord ice at the mouth of Kangia who fished with lines of baleen. Poul Egede measured a line, which was no less than 684 metres long.

During hard winters, when famine threatened, a large part of the population living in the Disko Bugt region would make long journeys to the icefjord to fish for halibut. B.J. Schultz, the colonial inspector of Qeqertarsuaq, described the situation in the hard winter of 1792-1793. Hungry Greenlanders, who lacked dogs, walked the 150-225 km from Qeqertarssuaq and Aasiaat over the sea ice to Kangia to reach the halibut fishing grounds. Similar journeys were also recorded during later periods of famine.

some people believed the moon had taken her, while others more prosaically thought that her husband had got rid of her by pushing her under the ice.

During his visits, Poul Egede told them about Christianity, and the people of Eqi were reportedly eager to hear the word of God. Egede describes a large sacrificial stone, which some boys showed him. Sacrifices were made by spreading blubber on the stone, and placing sealskin and fish in cracks in the stone. The sacrifices were made in order to secure a good catch, and the stone was still an object of heathen-worship as late as 1778.

Unfortunately Poul Egede says nothing about how many houses there were at Eqi, nor how many people lived there, but just reports that the place was populous. Available records indicate that in 1811 a total of 40 people lived in the settlement, in 1915 there were 33 people and in 1933 there were 57 inhabitants. Now nobody lives at the icefjord's 'southern corner of the mouth', the English translation of Eqi.

Colonisation and the Napoleonic Wars (1800-1830)

The colony of Jakobshavn (now Ilulissat) was established in an area where there were already a number of settlements. The Inuit population lived from traditional hunting, but was naturally attracted by goods such as the firearms, hardware and pearls available in the colony. Barter was widespread outside the regime of 'Kongelige Grønlandske Handel' (KGH - The Royal Greenland Trading Company). The colony Ritenbenk (now Appat) was established north of Jakobshavn in 1755.

Inuit at their winter house. After the Danish colonisation of 1721, it became more common for the Inuit to use their winter houses throughout the year.

In 1805 there were 86 people living in Jakobshavn, but about twice as many still lived in Sermermiut and the other settlements around the colony.
The people of Sermermiut slowly started to move from the turf houses of their settlement to Jakobshavn, in part because of improved possibilities of trade with Europeans.

This was towards the end of the great whaling period, when blubber was rendered into whale oil, and provided the fuel to light the lamps of Europe. However, whales were close to extermination in the area and the whaling installations at Jakobshavn closed down in 1800. During the Napoleonic Wars supplies from Denmark were limited, and the time from 1807 to 1814 in particular was a

FISHING FROM A BOAT OFF THE ICEFJORD

period of great hardship. In Jakobshavn, the hunters were forced to re-melt the lead from the roof and windows of the Zion Church in order to make rifle bullets. At the same time the British took over the Dutch role of illegal whaling and trading, and thus became competitors to the Danish trade monopoly in Greenland.

Recovery, whale oil and education (1830-1850)

The 1830s were characterised by contemporary thoughts of free trade and free competition as well as the start of industrialisation. It became a time of recovery, when production was improved, and increasing numbers of people moved to Jakobshavn from the surrounding settlements. In fact there were so many new arrivals that the colonial administrators were forced to attempt to counteract the centralisation tendency. This was mainly because the hunting focus was now increasingly on catching seals rather than the scarce whales, and sealing was most effective when the hunters hunted along the whole length of the coast. To encourage the wider distribution of hunting, the trading station Paakitsoq was established north of Ilulissat in 1832, although Paakitsoq never had any great importance.

The good times continued, and in the 1840s Jakobshavn could boast of having the third-largest production of blubber in the region. The large trade of seals had, however, a negative aspect: in their eagerness to buy new, cheap goods, the Greenlanders sold far more seal skins and blubber than they needed to maintain their means of existence. The result was empty winter larders, hunger and a lack of clothing, tents and boats. The trade records for Jakobshavn in 1844 show that of 809 seals caught, the skins from 798 were sold. The 300 people in the colony thus had only 11 skins left for their own use.

In 1848 the training college at Jakobshavn was built and, together with the college at Godthåb (now Nuuk), these were the first higher education establishments in Greenland. The catechists trained at the colleges created a cultural advance that gave the Greenlanders a special position in relation to the Inuit populations elsewhere in the Arctic. However, the college in Jakobshavn closed in 1907 due to declining numbers of pupils, and today the building is a museum.

The exodus from Sermermiut and the tuberculosis epidemic (1850-1890)

The excessive sale of skins and blubber vital to their way of life had serious consequences for the inhabitants of Sermermiut. Their reserves were exhausted, and since the catch fell at the same time, many were plagued by hunger and epidemics, which were more serious because of poor housing conditions.

The last inhabitants of Sermermiut moved to Jakobshavn in 1850, bringing the population up to 262. Thirty years later the population had increased to 418. In a new attempt at attracting people away from Jakobshavn, the trading station Ataa, north of Jakobshavn, was created successfully in 1852. A blubber house was built and trade in blubber and shark livers was started. In 1939, 35 people still lived there, but it has now been abandoned.

Foreign ships called regularly at the colony and the trading stations and, unfortunately, the shipping traffic brought more than commodities. In 1867 there was an outbreak of tuberculosis in Jakobshavn, and a third of the population were carried to the cemetery. Such large losses naturally affected fishing and hunting, and it was difficult to keep the small community running.

Three of Jakobshavn's great personalities were born in these difficult times. The catechist Jørgen Brønlund entered the world in 1877; he participated in several polar expeditions, but lost his life during the Danmark Expedition in 1907. Knud Rasmussen, the renowned arctic explorer, was born in 1879 in the former college building. He died in 1933 of an illness contracted during the 7th Thule Expedition. Mathias Storch was born in 1883 near Aasiaat; he was trained and worked as a priest in Jakobshavn and became the first Greenlander to be appointed viceprovst (rural dean) in 1928. Mathias Storch later wrote the first Greenlandic novel and was particularly active in politics and social debate.

Halibut for export (1890-1930)

By the 1890s, fishing for Greenland halibut and the Greenland shark had become the main occupation of the population of Jabobshavn and the nearby trading stations. The Greenland shark was an important source of income, as its liver was sold to The Royal Greenland Trading Company (KGH), while halibut was then only eaten and traded locally. It is worth noting that the communities around the icefjord became skilled fishermen while the rest of Greenland's population primarily depended on catching marine mammals for income and food.

Photo: Dieter Zillmann/Greenland Tours Elke Meissner

Long-line fishing from the sea ice in the icefjord. The hole with the line is in the foreground..

Halibut acquired new significance for the local economy about 1900. The colony managers, or 'factors', had started to export a few hundred kilograms of halibut fillets to their friends and families in Denmark, which they were allowed to do as halibut was not covered by the KGH monopoly. The private export grew markedly after Poul Müller, the colony manager at Jakobshavn from 1892 to 1901, established contact with smokehouses

THE BEUTIFUL LOCATION OF THE ZION CHURCH

in Copenhagen. Following his initiative, the landings of halibut in Jakobshavn increased to more than 22 tons in 1899.

Then KGH imposed strict restrictions on the private export of halibut, and actually took over the trade in 1903. Less than two tons were exported in the early years under the KGH monopoly, but the export grew quickly again, and halibut exports have since become vital for the economy of Jakobshavn (now Ilulissat). Halibut fishing saw further advances at the end of the 1920s, when Marius Sivertsen invented a so-called 'glider' that revolutionised long line fishing at the icefjord.

The change from hunting to fishing halibut that occurred in the 1890s had a marked impact on the pattern of settlement in the district. The authorities had to reverse the decision to spread the population to encourage hunting, since the new developing main trade required a concentration of the fishermen where there were fish and fish processing possibilities. At the same time, the concentration on fishing meant that the population changed from barter to a money economy, as fish do not give the same possibility for self-sufficiency as sealing.

The period was an active one for new construction and enterprise. In 1924, the coal mine Qullissat was opened on Disko island, and the mine became a major employer that attracted labour from the whole of the Disko Bugt region. The production of coal at Qullissat was so large that it satisfied most of Greenland's needs, but competition from cheaper British coal imports led to the mine's closure just under 50 years later in 1972.

The first part of the present hospital in Ilulissat was built in 1926, and it was decided to move the Zion Church 50 m away from the sea in 1929-1932, presumably to avoid influence from the sea.

World war, reform and centralisation (1930-1960)

Denmark had anticipated the Second World War was coming, and extra supplies were sent to Greenland as early as the summer of 1939. When the war broke out, it soon became clear that Greenland could not do without support from abroad. A direct diplomatic connection was therefore negotiated with the USA, which took over the responsibility for supplying Greenland with necessities. The war had great importance for Greenland, which experienced the period as a time of prosperity. At the same time the disparity between Greenland and the outside world became obvious as the Americans overcame the problems of the arctic environment to build military bases, and introduced modern consumer goods.

After the war there came a time of change. The comprehensive complex of laws, 'Nyordningen', made Greenland an integral part of Denmark in 1953. One aspect of 'the reform' aimed at concentrating the population geographically, so that fishing could be intensified. At the same time it was decided that the traditional primitive houses in which most of the Greenlanders lived would be replaced by modern houses, and the door was

Photo: The Royal Library

Knud Rasmussen was the most important bridge-builder between the Danish and Greenlandic cultures.

Knud Rasmussen, also called 'Little Knud', was born in Jakobshavn (Ilulissat) in 1879, the son of a Danish father and a Danish-Greenlandic mother. The Inuit language and culture fascinated Rasmussen, and he spent most of his life on ethnographic studies of the Inuit in Greenland, Canada and Alaska.

In the years 1902-1904 Knud Rasmussen embarked on his first major expedition, together with Jørgen Brønlund, Elias Eliasen, Gaba Olsen, Harald Moltke and Ludvig Mylius-Erichsen. In 1910 Rasmussen established the Thule trading station together with Peter Freuchen, and his subsequent expeditions were known as 'Thule Expeditions'. On his journeys, Knud Rasmussen recorded countless myths and sagas, as well as collecting examples of clothing and tools.

From 1921 to 1924 Knud Rasmussen led the Fifth Thule Expedition, the most celebrated of the seven Thule expeditions, when he undertook a 46,000 km sledge journey across northern Canada to Alaska. This sledge journey was of fundamental importance to the understanding of the links between the different Inuit cultures of the Arctic. In recognition of his achievements, Knud Rasmussen was awarded honorary doctorates by the universities of Copenhagen and Edinburgh.

Knud Rasmussen's Danish-Greenlandic background provided him with the insights and ability to live and think like the Inuit he visited. He was the most important bridge-builder between the Danish and Greenlandic cultures. When Knud Rasmussen died in 1933 there was public mourning in both Greenland and Denmark.

The birth place of Knud Rasmussen is now a museum.

From the photo archive at the Arctic Institute, Copenhagen

The participants in the Danish 'Litterære' expedition 1902-04. From left Knud Rasmussen, Ludvig Mylius-Erichsen, Osaqaq and Jørgen Brønlund.

Mylius-Erichsen and Jørgen Brønlund perished in 1907 together with Høeg Hagen during the Danmark expedition.

Photo: © The National Museum of Denmark, Etnographical Division

Rasmussen's many great sledge journeys through Greenland Arctic Canada and Alaska, made him world famous.

Photo: Dieter Zillmann/Greenland Tours Elke Meissner

FISHING BOATS IN A FANTASTIC SCENARIO

opened for private initiative, although still under state control.

'Nyordningen' was the trigger for the greatest technological development Greenland has ever seen, and ten years later the country was almost unrecognisable. An effective infrastructure, numerous new installations and modern facilities were turned into reality at great speed. The reform awoke hopes in the community for equality between Greenland and Denmark, but those hopes were not realised despite the great progress - partly because Denmark in the same period went through an larger development, an partly because there were great differences in pay conditions between Denmark and Greenland.

The influx of large numbers of cod to West Greenland in the period after the war led to development of the cod fishery. In Disko Bugt the cod fishing led to an expansion of facilities at Jakobshavn. The children's home was built in 1953-1954, the quay to service ships crossing the Atlantic to fish and trade came in 1960, and the fish-processing factory and the Vandsø dam in 1961.

Shrimp bonanza, population explosion and tourism (1960-2000)

The cod eldorado did not last long. There was a general reduction in cod stocks along the coast of West Greenland in the 1960s, but, very fortunately, a new and even more profitable catch had been discovered - enormous stocks of shrimps. In Ilulissat people adapted quickly, the necessary production installations were established and the new business attracted many migrants from the outlying villages. By 1974 there were nearly 1500 people in the town, while the population 20 years later had risen to 3800.

The intense economic development of the fishing industry, the centralisation and the improved living conditions resulted in something close to a population explosion, which was further strengthened when, in the middle of the 1970s, large numbers of workers from Denmark were recruited to Greenland. Many of the modern facilities in Ilulissat date from that time. The town's two large schools were established in 1974 and 1977, the Hotel Hvide Falk is from 1971 and the sports hall was added in 1976. At the same time several large housing complexes were built. The hospital was extended in 1980.

As noted above, the coal mine at Qullissat on Disko closed in 1972, and many of the former workers moved to Ilulissat. The new area of the town where they were housed acquired the name 'Little Qullissat'; this area was close to the present road known as Mathias Storchsvej. The closure of Qullissat, and the forced movement of so many workers, created much unrest, and became a symbol of how the authorities in far-away Denmark made decisions over the heads of the Greenlanders.

In 1979, Greenland acquired home rule following a referendum, turning Greenland into an autonomous part of the Kingdom of Denmark on equal footing with the Faeroe Islands. Home rule means increased responsibility in terms of self-administration and public matters. However, Denmark continues to provide Greenland with substantial economic subsidies via the so-called block grants, which are of the same size as before the introduction of home rule. With home rule, the Greenlanders acquired a new self respect, and one result of this was the deliberate policy to abandon the Danish names formerly used in conjunction with the Greenlandic names for towns. Jakobshavn became Ilulissat.

Tourism came to Greenland in a big way in the 1980s, and especially to Ilulissat, which is today the country's most-visited tourist town. Annually Ilulissat has more than 10,000 tourists. Ilulissat puts a lot of effort into tourism, especially in providing opportunities for fishing. Other forms of tourism are limited by the expense of air flights, a very short tourist season and the limited hotel capacity.

Hunting mammals, fish and birds

Exploiting animals for food and clothing has always been the foundation for man's existence in Greenland, and before colonisation the population was completely dependent on hunting mammals, fish and birds. Various species were hunted at different times of the year, and the Inuit moved from place to place in the course of the year.

Following colonisation, rifles and other modern tools made hunting more effective. At the same time, access to European food in exchange for skins, blubber and tusks, meant that the nomadic lifestyle became less important. As a result the Inuit tended to concentrate around the colonial stations, and these living sites became progressively more permanent. Remnants of the former nomadic lifestyle are still seen today, when hunting families live in tents in the summer, and stay at arctic char streams or hunt reindeer for a few weeks.

Ready for the winter season.

Many of the traditional hunting weapons and forms of transport have now been replaced by rifles, small boats and nylon net, but hunting and fishing nevertheless

FISH ARE BEING DRIED FOR THE WINTER

support the traditional hunter culture. In the Ilulissat area, the dog sledge is still widely used during the winter when snow and ice covers the area.

Some hunted species have declined markedly as a result of the efficiency of modern hunting methods. To counteract the tendency, politicians have introduced laws and procedures that regulate hunting. Hunters today must take out a hunting licence, which can be granted to anyone over the age of 12. Licences are issued either for what are termed recreational hunters or to professional hunters.

Recreational hunting

Recreational hunting has existed since the first Greenlanders were employed by the colonial administration. In Ilulissat, the prey of the recreational hunters includes seals, ptarmigan, arctic hares, foxes, sea birds and fish, as well as musk ox and reindeer if the hunters are willing to travel long distances. When the water is open, boats are mainly used for transport, while dog sledges are used for transport on ice in winter.

In the winter of 2001-2002 there were about 10 registered recreational hunters, who hunted ringed seals using nets under the ice. In October, some hunters harpoon seals coming to their breathing hole in the ice, and in spring the seal is shot while it is lying in the sun on the ice. A small sled, with a white sail as camouflage, is used to approach the prey to within firing distance.

It takes from four to six hours to travel from Ilulissat to Sikuiuitsoq, north of the icefjord, using a dog sledge, but many recreational hunters make the trip almost every weekend in the sledging season, which indicates how important hunting is for them. If one asks why, the answer is always 'economics'. This reply is understandable when one takes into account how much meat and fish a family of five people with ten dogs eats - about 500 kg per month.

The importance of recreational hunting for the hunter and his family with respect to social status, cultural awareness and maintaining the traditional lifestyle is difficult to assess. However, it is evident that hunters who use traditional methods of transport and hunting techniques, and who can share meat from the catch with friends and family, are an important part of the Greenlandic identity.

Professional hunting

At Ilulissat, most of the professional hunters' income comes from catching halibut. When there is not too much ice, boats and fishing vessels from Ilulissat, Ilimanaq, Oqaatsut and Qasigiannguit all sail to the mouth of Kangia to fish.

In winter, most of the fishing takes place from the sea ice, where dog sledges are the only form of transport. Fishermen with dog sledges from Qasigiannguit fish in the southern part of the icefjord, east of Qajaa, while fishermen from Ilulissat usually fish in the northern parts of the fjord. From 60 to 100 professional fishermen with dog sledges were active in the Ilulissat area in the winter of 2001-2002.

Two fishermen on the sea ice of Kangia, with some halibut on the dog sledge.

Photo: Dieter Zillmann/Greenland Tours Elke Meissner

The region around Kangia, showing main dog sledge routes (purple lines), planned huts (purple circles), and important fishing areas (broad pink lines and areas).

LAKE NEAR THE ICEFJORD

TRADITIONAL HOUSES IN ILULISSAT

In addition to halibut, the catch can include ringed seals, geese, ducks, arctic hares and ptarmigan. Eggs are gathered from gull colonies at Sikuiuitsoq in the early summer, when ice conditions permit travel by dog sledge. Occasionally narwhals are caught in the outer parts of the icefjord. Reindeer are hunted southeast of the Ilulissat area in autumn. However, hunting mammals and birds only plays a relatively small role compared to fishing for halibut.

Both the fishing areas and the hunting areas can change from year to year as a result of changes in the ice conditions and the snow cover. Over longer periods, such as decades or centuries, it is clear that fishing and hunting areas have moved significantly. Some of the present areas for hunting and fishing lie to the east of the place where the glacier front of Sermeq Kujalleq was located in 1850. The exploitation of the living resources in Kangia is a good example of how the people in the area have adapted to the ever-changing glacial environment.

Fishing methods, tackle and localities

Originally, halibut was fished using jigs of soapstone and bone, and with a fishing line of baleen from the Greenland right whale. At the beginning of colonisation, use of iron hooks became common and the traditional fishing equipment was occasionally modified by replacing the baleen line with fishing lines of hemp. But baleen lines were used in preference to hemp lines as late as 1852, as it is easier to feel the fish biting with a baleen line when fishing for halibut.

The fishing banks lie at the mouth of Kangia, close to Ilulissat and the former trading stations of Pitoqqeq, Illumiut and Kingittoq. Under normal circumstances it is not necessary to travel very far to reach the good fishing areas. However, in the spring and the autumn, when the sea ice in front of the houses is unsafe, the fishing area near Seqinniarfik is preferred, as the ice is usually thicker here.

In earlier times there were occasionally no halibut in the northern part of Kangia in winter, probably because of the migration of the belugas from the end of April to June and again from September to December. Belugas hunt halibut, and with their arrival the halibut migrated to other areas. When that happened, many hunters left the northern part of Kangia, and made the long trip south across the mouth of the icefjord to the fishing areas at Tasiusaq, behind the trading station Ilimanaq.

Long line fishing for halibut started in the fjord about 1906. The inspiration came from the fishermen of the Faeroe Islands, and the adoption of their long line method made the fishing much more effective. Marius Sivertsen's 1924 'glider' invention made long line fishing from the sea ice more effective in winter, and led to an expansion of the fishing area. In 1929, Nicolai Olsen and others from Jakobshavn made the first journey to the inner part of Kangia, and discovered the fishing banks at Nunatarsuaq.

This was the start of winter fishing in the icefjord itself. The fishing areas expanded to include Aallaaniarfik and Nalluarsuk, and these proved to be an important supplement to the traditional winter areas on the fjord ice between Seqinniarfik and Ilulissat. Use of the 'glider' for long line fishing has since spread to other parts of West Greenland. In this way, the fishermen at Kangia have played a central role in the development of fishing in Greenland.

Gill-nets have been used in both summer and winter fishing off the mouth of Kangia since the 1980s, but the quality of the halibut recovered is not as high as with long line fishing. The method is mostly used from fishing boats, but the loss of nets can cause 'ghost fishing' and there are therefore many restrictions on the use of gill-nets.

Photo: Henrik Højmark Thomsen

The mouth of the icefjord wish the fishing grounds near Ilulissat.

THE COMING GENERATION IS LISTENING

Greenlandic choir in national dress.

Within the boundaries of the Ilulissat Icefjord World Heritage site, travel in connection with winter fishing is presently by dog sledge. In periods with only little ice, fishing is carried out from boats with outboard motors, or from small fishing vessels at the mouth of the icefjord.

Royal Greenland A/S runs the fish factory in Ilulissat, which is the largest in Greenland and produces a range of products from halibut and shrimp for the international markets. Processing the halibut employs over 200 part-time workers, in addition to 20 full-time employees. Between 1997 and 2001 the annual value of the halibut sold by the fishermen to the factory was 30-45 million Danish kroner. Fishing for halibut, and the associated fishing industry, is one of the most important sources of employment and income in Ilulissat.

The fish factory uses all parts of the halibut, and this contributes to the clean environment in the area around Ilulissat. Fish offal, such as heads, intestines, fins, bones and skin are processed into fishmeal, and surplus fish oil in the offal is used for heating the large factory building and to produce steam for the shrimp boiling house. In addition, a large part of the fish offal is sold as dog food, which is eaten by the more than 4000 sledge dogs in the town.

Ilulissat today

For the past two decades, Ilulissat has continued its rapid development, and the town, which was founded as a modest colony at a natural harbour on the Greenland west coast, has grown large. Today, Ilulissat is the largest town in the Disko Bugt region, the third-largest town in Greenland, and a centre of education, culture, administration, fishing and tourism. The inhabitants are very active in the community, and have taken the initiative for many local projects. There are local TV and radio stations and a local newspaper, and dog-sledge trips, ski races, handball matches and musical performances are organised.

The number of inhabitants in Ilulissat and the four surrounding villages passed 5000 in 2003. It is a far cry from the turf houses and kayaks of Sermermiut to today's multi-story housing blocks, snowscooters and electronic media; but there are still people living in the town whose grandparents lived in Sermermiut. Ilulissat is thus an excellent example of an arctic 'city', where tradition in combination with new businesses has ensured the existence of an active and living town.

The standard of living has improved considerably in recent decades. The average life expectancy in Greenland is 65 years, which may be considered low compared with the western world, but people live on average ten years longer than they did just 50 years ago. The causes of the low average life expectancy are the high infant mortality rate, many deaths in accidents at work, and suicides.

Tourists admiring the icebergs.

Photo: Dieter Zillmann/Greenland Tours Elke Meissner

The days of an economy based on payment in kind are long gone, and the average income in 2001 was 106,989 Danish kroner. If one wishes to know something about a country's economic productivity, the gross national product per capita provides a good estimate. In Greenland this was 153,647 kroner in 2000, equal to 57 per cent of the gross national product per capita in Denmark, so there is still a large difference between the two parts of the kingdom.

A USEFUL ROAD SIGN IN ILULISSAT

Unemployment is low in Ilulissat, which has one of Greenland's lowest jobless rates - 4.3% in 2002 - and even if one is unemployed it is still possible to catch one's own food. Many receive help from the close family networks, where there is tradition for mutual assistance, e.g. as gifts of meat.

The standard of housing has improved considerably, but there is still a serious shortage of housing in Ilulissat. The available housing covered just over 76% of housing needs in 2003, which is considerably lower than in other Greenland towns. The solid Greenland bedrock means that laying water pipes, sewage pipes and district heating pipes is often difficult. Many houses still use water from tanks in their houses, which are filled from the town's water tanker, and there are still a very small number of the inhabitants who collect their water from the municipal water taps. The larger housing blocks in particular have running water, and here there are also flush toilets. However, it is not unusual for modern houses to have earth closets, which are emptied by the local authorities. The use of district heating is spreading, again mainly in the larger housing blocks, although many people heat their homes with oil or paraffin heaters.

The four villages in the municipality of Ilulissat - Saqqaq, Qeqertaq, Oqaatsut and Ilimanaq - do comparatively well.

Ilulissat's coat of arms.

There are fish production installations at Qeqertaq and Oqaatsut, but the population is still dependent on both hunting and social support. Nevertheless, the four villages are collectively among the few villages in Greenland where the population is rising. A number of tourists visit Oqaatsut, which is frequently named as an example of what village life has to offer tourists. Ilimanaq is also doing fairly well in respect of tourism and there are expectations that the Ilulissat Icefjord World Heritage site will attract tourists to the whole of the beautiful Disko Bugt region.

Ilulissat municipality is led by a council of 13 elected members. Under the municipal council there are a number of committees with specific tasks. In addition there is a town administration. The municipality runs a large number of institutions of different kinds, such as crèches, kindergartens, schools, a music school, educational centres, juvenile recreational centres, meeting places, a library and a retirement home. Several institutions in the town are run by the Greenland Home Rule administration, including the training college for 'socialpædagoger', the hospital and the power station. The Danish state is responsible for the administration of justice, the prison service and the police.

Fishing is the main occupation in Ilulissat, and there is a large fleet of shrimp trawlers, and smaller vessels and boats for fishing, primarily for halibut. About a quarter of the town's population works at the Royal Greenland fish factory.

Ilulissat is Greenland's leading tourist town and about a half of all the tourists who visit Greenland come to Ilulissat. The icefjord Kangia, in particular, is the attraction, and the easy access by air provided by the airport is very important. There are good possibilities for excursions by dog sledge, boat and helicopter, and for hiking trips. One can visit the nearby villages, and in Ilulissat one can stay at fine, four-star hotels, a more modest youth hostel or at the camp site.

The local history museum is worth seeing. It is housed in the old college building and contains among other things an exhibition about the famous arctic explorer Knud Rasmussen. Outside the museum there is a copy of a traditional winter house, an old whale oil works and two lower jaws from a Greenland right whale.

Ilulissat has been a traffic hub for the Disko Bugt region since the airport was built in 1984. From the airport there are connections to, amongst other places, Qaanaaq (Thule) in the north and Kangerlussuaq in the south - and thus Europe. There are also connections to Canada and North America. The world's largest dog sledge is to be found at the airport; 23 metres long and pulled by 260 dogs it made a 110-metre trip in 1994 with no less than 108 passengers. There is also a large sleigh that was built for the Santa Claus World Congress held in Ilulissat in 1995.

Ilulissat is a modern Greenland town with many facilities, and has ambitions. There are proposals to expand the airport, for example, so that larger aircraft can fly direct to the town, which would benefit both tourism and fish exports. That Ilulissat Icefjord, centred on the unique and majestic splendour of the icefjord south of Ilullisat, has now been added to UNESCO's World Heritage list will be of vital importance for the town's future development.

The airport was inaugurated in 1984.

Photo: Ivars Silis

Supplementary reading

THE SLEDGE DOGS ARE OUT OF WORK DURING THE SUMMER

A comprehensive reading list can be found on the Internet at: http://www.geus.dk/publications/boeger/ilulissat-uk.htm

Alley, R. 2000: The two-mile time machine: ice cores, abrupt climate change, and our future. 229 sider. Princeton: Princeton University Press.

Amdrup, G.C., Bobé, L., Jensen, A.S. & Steensby, H.P. (red.) 1921: Grønland i Tohundredåret for Hans Egedes Landing. Meddelelser om Grønland 60-61. 567 sider + 795 sider + atlas.

Bauer, A., Baussart, M., Carbonnell, M., Kasser, P. Perroud, P. & Renaud, A. 1968: Missions aériennes de reconnaissance au Groenland 1957-1958. Observations aériennes et terrestres, exploitation des photographies aériennes, determination des vitesses des gletschers vélant dans Disko Bugt et Umanak Fjord. Meddelelser om Grønland 173, 3, 116 pages.

Bendixen, O. 1930: Grønlandsfiskeriet, dets Historie og Fremtidsmuligheder. 129 pages. Copenhagen: H. Aschehoug & Co.Boertmann, D. 2002: Fugle i Grønland. 80 pages. Nuuk: Ilinniusiorfik Undervisningsmiddelforlag.

Böcher, T. 2000: Det grønne Grønland. 2. edition. 432 pages. Copenhagen: Rhodos.

Böcher, T.W., Fredskild, B., Holmen, K. & Jakobsen, K. 1978: Grønlands flora. 326 pages. Copenhagen: P. Haase & Søn.

Boertmann, D. 2002: Fugle i Grønland. 80 pages. Nuuk: Ilinniusiorfik Undervisningsmiddelforlag.

Born, E.W. & Böcher, J. (red.) 1998: Grønlands Økologi - en grundbog. 431 pages. Nuuk: Atuakkiorfik Undervisning.

Carbonnell, M. & Bauer, A. 1968: Exploitation des couvertures photographiques aériennes répétées du front des gletschers vélant dans Disko Bugt et Umanak Fjord, juin-juillet 1964. Meddelelser om Grønland 173, 5, 78 pages.

Clarke, T.S. & Echelmeyer, K. 1996: Seismic-reflection evidence for a deep subglacial trough beneath Jakobshavns Isbræ, West Greenland. Journal of Glaciology 42, 219-231.

Cranz, D. 1770: Historie von Grönland enthaltend die Beschreibung des Landes und der Einwohner, 2. edition. 710 pages. Barby: Heinrich Detlef Ebers.

Dansgaard, W. 1987: Klima, vejr og menneske. 128 pages. Brenderup: Geografforlaget.

Dansgaard, W., Johnsen, S.J., Clausen, H.B., Dahl-Jensen, D., Gundestrup, C.U., Hvidberg, C.S., Steffensen, J.-P., Sveinsbjørnsdottir, A.E., Jouzel, J. & Bond, G. 1993: Evidence for general instability of past climate from a 250 kyr ice core record. Nature 364, 218-220.

Echelmeyer, K. & Harrison, W. D. 1990: Jakobshavn Isbræ, West Greenland: Seasonal variations in velocity - or lack thereof. Journal of Glaciology 36, 82-88.

Echelmeyer, K., Clarke, T. S. & Harrison, W. D. 1991: Surficial glaciology of Jakobshavn Isbræ, West Greenland: Part I, Surface morphology. Journal of Glaciology 37, 368-382.

Echelmeyer, K., Harrison, W. D., Clarke, T. S., & Benson, C. 1992. Surficial glaciology of Jakobshavn Isbræ. Part II, Ablation, accumulation and temperature. Journal of Glaciology 38, 169-181.

Egede, P. 1788: Poul Egedes Dagbog. Efterretninger om Grønland, uddragne af en journal holden fra 1721 til 1788. Edited by Mads Lidegaard. Det Grønlandske Selskabs Skrifter 29, 204 pages.

Engell, M. C. 1904: Undersøgelser og Opmålinger ved Jakobshavns Isfjord og i Orpigsuit i Sommeren 1902. Meddelelser om Grønland 26, 1, 1-70.

Engell, M. C. 1910: Beretning om Undersøgelserne af Jakobshavns Isfjord og dens Omgivelser fra Foråret 1903 til Efteråret 1904. Meddelelser om Grønland 34, 6, 155-251.

Epprecht, W. 1987: A major calving event of Jakobshavn Isbræ, West Greenland, on 9th August 1982. Journal of Glaciology 33, 160-172.

Escher, J.C. & Pulvertaft, T.C.R. 1995: Geological map of Greenland, 1:2,500,000. Copenhagen: Grønlands Geologiske Undersøgelse.

Fastokk, J.L., Brecher, H.H. & Hughes, T.J. 1995: Derived bedrock elevations, strain rates and stresses from measured surface elevations and velocities: Jakobshavn Gletscher, Greenland. Journal of Glaciology 4, 161-173.

Feilberg, J., Fredskild, B. & Holt, S. 1984: Grønlands blomster. 98 pages. Ringsted: Ahrent Flensborg.

Fisker, J. 1984: Ilulissat Jakobshavn. 303 pages. Uummannak: Nordiske Landes Bogforlag.

Fleischer, J. 2003: Grønlands historie - kort fortalt, 70 pages. Copenhagen: Aschehoug.

Fredskild, B. 1961: Floristic and ecological studies near Jakobshavn, West Greenland. Meddelelser om Grønland 163, 4, 82 pages.

Fredskild, B. 1967: Palæobotanical Investigations at Sermermiut, Jakobshavn, West Greenland. Meddelelser om Grønland 178, 4, 54 pages.

Fristrup, B. 1966: Grønlands Indlandis. 319 pages. Copenhagen: Rhodos.

Friis, Achton. 1987: Danmark Ekspeditionen. 431 pages. Copenhagen. Gyldendal.

Funk, K., Echelmeyer, K. & Iken, A. 1994: Mechanisms of fast flow in Jakobshavn Isbræ, West Greenland, Part II, modelling of englacial temperatures. Journal of Glaciology 40, 569-585.

Gad, F. 1969-1978: Grønlands historie. Vol. 1 (1978): Indtil 1700, 461 pages. (Oprindeligt publiceret i 1967). Vol. 2 (1969): 1700-1782, 760 pages. Vol. 3 (1976): 1782-1808., 615 pages. Copenhagen: Nyt Nordisk Forlag, Arnold Busck.

Gensbøl, B. 1998: Grønlands dyr & planter. 204 pages. Copenhagen: Gad.

Georgi, J. 1960: Der Rückgang des Jakobshavns Isbræ (West-Grönland 69° N). Meddelelser om Grønland 158, 5, 51-70.

Grønnow, B. 1994: Qeqertasussuk - the Archaeology of a Frozen Saqqaq Site in Disko Bugt, West Greenland. In: Morrison, D. & Pillon, J.-L. (eds.): Threads of Arctic Prehistory: Papers in Honour of William E. Taylor, Jr. Canadian Museum of Civilization, Mercury Series. Archaeological Survey of Canada Paper 149, 197-238.

Grønnow, B. 1996: The Saqqaq Tool Kit - Technological and Chronological Evidence from Qeqertasussuk, Disko Bugt. In: Grønnow, B. & Pind, J. (eds.): The Paleo-Eskimo Cultures of Greenland - New Perspectives in Greenlandic Archaeology. Copenhagen: Danish Polar Center Publication No. 1/1996, 17-34.

Grønnow, B. & Meldgård, M. 1988: Boplads i dybfrost - fra Christianshåb Museums udgravninger på Vestgrønlands ældste boplads. Naturens Verden 1988, 409-440.

Grønnow, B. & Meldgård, M. 1991: De første vestgrønlændere. Grønland 4-7, 103-144.

Gulløv, H.C. 1997: From Middle Ages to Colonial Times. Archaeological and ethnohistorical studies of the Thule culture in South West Greenland 1300-1800 AD. Meddelelser om Grønland, Man & Society 23, 501 pages.

Hammer, R.R.J. 1883: Undersøgelser ved Jakobshavns Isfjord og nærmeste Omegn i Vinteren 1879-1880. Meddelelser om Grønland 4, 1, 1-68.

Hansen, E.S. 1995: Grønlands Laver. 127 pages. Nuuk: Atuagkat and Copenhagen: Rhodos.

Hansen, P.M. & Hermann, F. 1953: Fisken og havet ved Grønland. Skrifter fra Danmarks Fiskeri- og Havundersøgelser 15, 128 pages.

Helland, A. 1876: Om de isfyldte Fjorde og de glaciale Dannelser i Nordgrønland. Archiv for Matematik og Naturvidenskab 1, 69 pages.

Hertz, O. 1995: Økologi og levevilkår i Arktis. 215 pages. Copenhagen: Christian Ejlers Forlag.

Huybrechts, P. 1994: The present evolution of the Greenland ice sheet; an assessment by Modelling. Palaeogeography, Palaeoclimatology, Palaeoecology (Global Planetary Section) 9, 39-51.

Huybrechts, P. 2002: Sea-level chances at the LGM from ice-dynamic reconstructions of the Greenland and Antarctic ice sheets during the glacial cycles. Quaternary Science Reviews. 21, 203-231.

Jakobsen, B.H., Bøcher, J., Nielsen, N., Guttesen, R., Humlum, O. & Jensen, E. (eds.) 2000: Topografisk Atlas, Grønland. 278 pages. Copenhagen: Det Kongelige Danske Geografiske Selskab & Kort & Matrikelstyrelsen.

Joughin, I., Abdalati, W. & Fahnestock, M. 2004: Large fluctuations in speed on Greenland´s Jakobshavn Isbræ glacier. Nature 432, 608-610.

Knuth, E. 1967: The Ruins of the Musk-ox Way. Folk 8-9, 191-219.

Koch, J.P. & Wegener, A. 1930: Wissenschaftliche Ergebnisse der dänischen Expedition nach Dronning Louises Land und quer über das Inlandeis von Nordgrönland 1912-13. Meddelelser om Grønland 75, 676 pages.

Kollmeyer, R.C. 1980: West Greenland outlet gletschers: An inventory of the major iceberg producers. International Association of Hydrological Sciences, Publication 126, 57-65.

Krabill, W., Abdalati, W., Frederick, E., Manizade, S., Martin, C., Sonntag, J., Swift, R., Thomas, R., Wright, W. & Yungel, J. 2000: Greenland ice sheet: high elevation balance and peripheral thinning. Science 289, 428-430.

Larsen, H. & Meldgård, J. 1958: Paleo-Eskimo Cultures in Disko Bugt, West Greenland. Meddelelser om Grønland 161, 2, 75 pages.

Laursen, D. 1950: The Stratigraphy of the marine Quaternary Deposits in West Greenland. Meddelelser om Grønland 151, 1, 142 pages.

Lorentzen, J. (ed.) 1999: Inuit, kultur og samfund - en grundbog i eskimologi, 256 pages. Århus: Systime.

Madsen, L.D. 2000: Fiskeriafhængighed, entreprenørskab og konsumfisk: Kolonibestyrer Poul Müllers eksport af hellefisk fra Ilulissat/Jakobshavn 1892-1901. Publikationer fra Institut for Geografi og Internationale Udviklingsstudier, Roskilde Universitetscenter 160, 145 pages.

Marquardt, O., Holm, P. & Starkey, D.J. (eds.) 1999: From sealing to fishing: social and Economic Change in Greenland, 1850-1940. Studia Atlantica 4, 106 pages.

Mathiassen, T. 1958: The Sermermiut Excavations 1955. Meddelelser om Grønland 161, 3, 52 pages.

Typical Greenlandic houses from the Disko Bugt area.

THE GLACIER FRONT IS INACCESSIBLE

Maxwell, M.S. 1985: Prehistory of Eastern Arctic. 327 pages. Orlando: Academic Press.

Meldgård, J. 1991: Bopladsen Qajaa i Jakobshavn Isfjord. Grønland 4-5-6-7, 191-205.

Mikkelsen, N. & Ingerslev, T. 2002: Nomination of the Ilulissat Icefjord for inclusion in the World Heritage List. 136 pages. Copenhagen: Danmarks og Grønlands Geologiske Undersøgelse.

Muus, B., Salomonsen, F. & Vibe, C. 1990: Grønlands fauna. 463 pages. Copenhagen: Gyldendal.

Møbjerg, T. 1983: An Ethno-archaeological Investigation of the Sermermiut Settlement, West Greenland. Folk 25, 23-50.

Møbjerg, T. 1986: A Contribution to Paleoeskimo Archaeology in Greenland. Arctic Anthropology 23, 19-56.

Møller, P.F. & Nielsen, S.E. 2003: Turen går til Grønland. 180 pages. Copenhagen: Politikens Forlag.

Nordenskiöld, A.E. 1870: Redogörelse før 1870 års expedition til Grönland. Öfversigt Kungliga Vetenskabs-Akademiens Förhandlingar 1870, 10, 973-1082.

Nørrevang, A. & Lundø, J. 1981: Grønlands natur. 567 pages. Copenhagen: Politikens Forlag.

Ostermann, H. 1941: Træk af Kolonien Jakobshavns Historie gennem 200 År, 1741-1941. Det grønlandske Selskabs Skrifter 11, 171 pages.

PARCA investigators 2001: Mass balance of the Greenland ice Sheet (PARCA). Journal of Geophysical Research 106, D24, 33689-34058.

Rabot, C. 1891: Une Excursion au Groenland. Le gletscher de Jakobshavn. Tour du Monde 31, II, 305-320.

Petersen, H.C. & Olsen, H.K. 1992: Nuna angallavigisarput. 117 pages. Copenhagen: Grønlands Geologiske Undersøgelse.

Reeh, N., Oerter, H., Lettréguilly, A., Miller, H. & Hubberten, H.W. 1991: A new, detailed ice-age oxygen-18 record from the ice sheet margin in central West Greenland. Palaeogeography, Palaeoclimatology, Palaeoecology (Global and Planetary Change Section) 90, 373-383.

Rignot, E. & Kanagaratnam, P. 2006: Changes in the velocity structure of the Greenland ice sheet. Science 311, 986-990structure of the Greenland ice sheet. Science 311, 986-990

Rink, H. 1853: Om den geographiske Beskaffenhed af de danske Handelsdistrikter i Nordgrønland, tilligemend en Udsigt over Nordgrønlands Geognosi. Det Kongelige Danske Videnskaberns Selskabs Skrifter, 5. Række, naturvidenskabelige og mathematiske Afdeling 3, 37-70.

Rink, H. 1857: Grønland geografisk og statistisk beskrevet, vol. 1: Det nordre Inspectorat, 420 pages, vol. 2: Det søndre Inspectorat, 588 pages. Copenhagen: Andr. Fred. Høst.

Rink, H. 1877: Danish Greenland, Its People and its Products, 468 pages. London: Henry S. King.

Rink, H. 1888: Nogle Bemærkninger om Indlandsisen og Isfjeldenes Oprindelse. Meddelelser om Grønland 8, 273-279.

Smidt, E.L.B. 1969a: The halibut, Reinhardtius hippoglossoides (Walb.), biology and exploitation in Greenland waters. Meddelelser fra Danmarks Fiskeri- og Havundersøgelser, New series 6, 4, 79-147.

Smidt, E. 1969b: Hellefisken (I-II). Grønland 1969, 257-270 & 353-366.

Sohn, H.G., Jezek, K.C. & van der Veen, C.J. 1997b: Jakobshavn Gletscher, West Greenland: 30 years of space-borne observations. Geophysical Research Letters 25, 2699-2702.

Thomas, R.H., Krabill, E & Frederick, J.K. 1995: Thickening of Jakobshavn Isbræ, West Greenland, measured by airborne laser altimetry. Annals of Glaciology 21, 259-262.

Thomas, R.H., Atkins, T., Csato, B., Fahnstock. M., Gogineni, P., Kim, C. & Sonntag, J. 2000: Mass Balance of the Greenland ice Sheet at High Elevations. Science 289, 426-428.

Thomas, R.H. and PARCA Investigators 2001: Prandram for Arctic regional climate assessment (PARCA): goals, key findings, and future directions. Journal of Geophysical Research 106, D24, 33691-33705.

Thomsen, H.H. 1986: Mass balance measurements at the margin of the Indlandisen near Jakobshavn, West Greenland. Polarforschung 54, 37-41.

Thomsen, H.H., Thorning, L. & Olesen, O.B. 1989: Applied gletscher research for planning hydro-electric power, Ilulissat/Jakobshavn, West Greenland. Annals of Glaciology, 13, 257-261.

Van der Veen, C.J. 2001: Greenland ice sheet response to external forcing. Journal of Geophysical Research 106, D24, 34047-34058.

Wadhams, P. 2000: Ice in the Ocean, 351 pages. Amsterdam: Gordon and Breach Science Publishers.

Warren, C.R. & Hulton, N.R.J. 1990: Topographic and glaciological controls on Holocene ice-sheet margin dynamics, central West Greenland. Annals of Glaciology 14, 307-310.

Weidick, A. 1966: Jakobshavn Isbræ i Fortid og Nutid. Grønland 1966, 361-386.

Weidick, A. 1969: Investigations of the Holocene Deposits around Jakobshavns Isbræ, West Greenland. In: Péwé, T.L. (ed.): The Periglacial Environment, 249-261. Montreal: McGill-Queens University Press.

Weidick, A. 1992: Jakobshavn Isbræ area during the climatic optimum. Rapport Grønlands Geologiske Undersøgelse 155, 67-72.

Weidick, A. 1995: Greenland. I: Williams, R.S. & Ferrigno, J.G. (red.): Satellite Image Atlas of Gletschers of the World. U.S. Geological Survey, Professional Paper 1386-C, 141 pages.

Weidick, A., Oerter, H., Reeh, N., Thomsen, H.H. & Thorning, L. 1990: The recession of the Indlandsisen margin during the Holocene climatic optimum in the Jakobshavn Isfjord area of West Greenland. Palaeogeography, Palaeoclimatology, Palaeoecology (Global and Planetary Section) 82, 389-399.

Epilogue

The Ilulissat Icefjord World Heritage Site was officially recognised in the summer of 2004. The original nomination volume formed the basis for the present book. Grateful thanks are extended to those who contributed to the nomination volume and/or this book, including Claus Andreasen and Joel Berglund, Greenland National Museum & Archives (archaeology), David Boertmann, National Environmental Research Institute (birds and mammals), Jesper Boje, Danish Institute for Fisheries Research (life in the icefjord), Jon Feilberg, Biomedia (plant life), Naja Habermann, Ilulissat Museum (recent history and Ilulissat today), Eric Steen Hansen, Botanical Museum (lichens), Lars Dyrløv Madsen, Greenland Home Rule (fishery), Kirsten Strandgaard, Ilulissat Museum (faith and story-telling tradition, hunting), Niels Henriksen and Christoph Mayer, Geological Survey of Denmark and Greenland (geology and glaciology). In addition we have received help, assistance and material from Naja Abelsen, Christian Bay, Søren Bundgård, Birgit Eriksen, A.K. Higgins, Torsten Ingerslev, Hans Kapel, Mark F. Meier, Elke Meissner, Jørgen Meldgaard, Karl Petersen, Jens Rosing, Jørgen Peder Steffensen, Henrik Højmark Thomsen, TNT architects and Dieter Zillmann.

We also thank the Directorate for Environment and Nature and Ilulissat municipality for their fruitful co-operation.

The Geological Survey of Denmark and Greenland (GEUS) has made exhaustive attempts to trace the copyright holders of photographs used in this volume; copyright holders who have inadvertently not been acknowledged should contact GEUS.

The present book about the Ilulissat Icefjord World Heritage Site has been produced with financial support from the Danish Ministry of the Environment as part of the environmental support program Dancea - Danish Co-operation for Environment in the Arctic and Ilulissat municipality.

At Ilulissat the mountains are sailing.

THE ICEFJORD WITH ITS CHAOS OF SMALL AND LARGE ICEBERGS

NIGHT OVER THE ICEFJORD